Lectures in Mathematics
ETH Zürich
Department of Mathematics
Research Institute of Mathematics

Raghavan Narasimhan

Compact Riemann Surfaces

Springer Basel AG

Author's address:

Raghavan Narasimhan
Department of Mathematics
University of Chicago
Chicago, IL 60637
USA

A CIP catalogue record for this book is available from the Library of Congress, Washington D.C., USA

Deutsche Bibliothek Cataloging-in-Publication Data
Narasimhan, Raghavan:
Compact Riemann surfaces / Raghavan Narasimhan. -
Springer Basel AG, 1992
(Lectures in mathematics)
ISBN 978-3-7643-2742-2 ISBN 978-3-0348-8617-8 (eBook)
DOI 10.1007/978-3-0348-8617-8

First reprint 1996

Originally published by Birkhäuser Verlag in 1992
produced from chlorine-free pulp. TCF ∞

ISBN 978-3-7643-2742-2

9 8 7 6 5 4 3 2

Preface

These notes form the contents of a *Nachdiplomvorlesung* given at the Forschungsinstitut für Mathematik of the Eidgenössische Technische Hochschule, Zürich from November, 1984 to February, 1985. Prof. K. Chandrasekharan and Prof. Jürgen Moser have encouraged me to write them up for inclusion in the series, published by Birkhäuser, of notes of these courses at the ETH.

Dr. Albert Stadler produced detailed notes of the first part of this course, and very intelligible class-room notes of the rest. Without this work of Dr. Stadler, these notes would not have been written. While I have changed some things (such as the proof of the Serre duality theorem, here done entirely in the spirit of Serre's original paper), the present notes follow Dr. Stadler's fairly closely.

My original aim in giving the course was twofold. I wanted to present the basic theorems about the Jacobian from Riemann's own point of view. Given the Riemann–Roch theorem, if Riemann's methods are expressed in modern language, they differ very little (if at all) from the work of modern authors.

I had hoped to follow this with some of the extensive work relating theta functions and the geometry of algebraic curves to solutions of certain non-linear partial differential equations (in particular KdV and KP). Time did not permit pursuing this subject, and I have contented myself with a couple of references in §17. These references fail to cover much other important work (especially of M. Mulase) but I have not tried to do better because the literature is so extensive.

It is a great pleasure to express my thanks to the ETH for its hospitality, to Prof. J. Moser for his encouragement, and to Dr. A. Stadler for the enormous amount of work he undertook which made these notes easier to write. But special thanks are due to Prof. K. Chandrasekharan. But for him, I would not have been at the ETH, nor would these notes have been written without his advice and encouragement.

Chicago, August 1991 R. Narasimhan

Contents

1. Algebraic Functions

Let $F \in \mathbb{C}[x, y]$ be an irreducible polynomial in two variables (with complex coefficients). We assume that its degree in y is ≥ 1.

Recall that by the so-called Gauss lemma, if we identify $\mathbb{C}[x, y]$ with $\mathbb{C}[x][y]$, and if F is irreducible, it is also irreducible in $\mathbb{C}(x)[y]$, the polynomial ring over the field of rational functions in x. Moreover, $\mathbb{C}[x, y]$ is a factorial ring (i.e. a unique factorisation domain).

An algebraic function is, intuitively, "defined" by an equation $F(x, y) = 0$ (where F is irreducible in $\mathbb{C}[x, y]$).

To make this statement more precise, we begin with the following.

The implicit function theorem. *Let f be a holomorphic function of two complex variables x, y defined on $\{(x, y) \in \mathbb{C}^2 \mid |x| < r_1, |y| < r_2\}$, $r_1, r_2 > 0$. Assume that*

$$f(0,0) = 0\,, \quad \frac{\partial f}{\partial y}(0,0) \neq 0\,.$$

Then, there exist positive numbers $\varepsilon, \delta > 0$ such that for any $x \in D_\varepsilon = \{z \in \mathbb{C} \mid |z| < \varepsilon\}$, there is a unique solution $y(x)$ of the equation $f(x, y) = 0$ with $|y(x)| < \delta$. The function $x \mapsto y(x)$ is holomorphic on D_ε.

Proof. Since $\frac{\partial f}{\partial y}(0,0) \neq 0$, we can choose $\delta > 0$ such that $f(0, y) \neq 0$ for $0 < |y| \leq \delta$. Choose now $\varepsilon > 0$ such that $f(x, y) \neq 0$ for $|x| \leq \varepsilon$, $|y| = \delta$ (possible since f is non-zero on the compact set $\{0\} \times \{y \mid |y| = \delta\}$).

By the argument principle, if $|x| < \varepsilon$,

$$\frac{1}{2\pi i} \int\limits_{|y|=\delta} \left\{\frac{\partial f}{\partial y}(x, y) \;/\; f(x, y)\right\} dy$$

is an integer $n(x)$ equal to the number of zeros of the function $y \mapsto f(x, y)$ in $|y| < \delta$; by our choice of δ, $n(0) = 1$. On the other hand, since $f(x, y) \neq 0$ for $|x| \leq \varepsilon$, $|y| = \delta$, the integrand, and thus also the integral, is a continuous function of x for $|x| < \varepsilon$. Thus $n(x) = 1$ for $|x| < \varepsilon$, which means precisely that there is a unique zero $y(x)$ of $f(x, y)$ with $|y(x)| < \delta$.

That $x \mapsto y(x)$ is holomorphic follows from the formula

$$y(x) = \frac{1}{2\pi i} \int\limits_{|y|=\delta} y \frac{\frac{\partial f}{\partial y}(x, y)}{f(x, y)}\, dy$$

(which is an immediate consequence of the residue theorem).

Let $F(x,y) = a_0(x)y^n + a_1(x)y^{n-1} + \cdots + a_n(x) \in \mathbb{C}[x,y]$ be an irreducible polynomial with $n \geq 1$; the polynomials $a_0, \ldots, a_n \in \mathbb{C}[x]$ have no non-constant common factor since F is irreducible.

Lemma 1. *Let $a \in \mathbb{C}$ be such that $a_0(a) \neq 0$ and such that there is no $b \in \mathbb{C}$ with $F(a,b) = 0 = \frac{\partial F}{\partial y}(a,b)$. Then, there is $\varepsilon > 0$ and n holomorphic functions $y_1(x), \ldots, y_n(x)$ in the disc $\{x \in \mathbb{C} \mid |x-a| < \varepsilon\}$ with the following properties:*

(i) $y_i(x) \neq y_j(x')$ if $i \neq j$, $|x-a| < \varepsilon$, $|x'-a| < \varepsilon$; moreover

$$F\big(x, y_i(x)\big) \equiv 0 \quad \text{for} \quad |x-a| < \varepsilon\,, \quad i = 1, \ldots, n\,.$$

(ii) if $\eta \in \mathbb{C}$ and $F(x,\eta) = 0$, $|x-a| < \varepsilon$, then $\eta = y_i(x)$ for a unique i between 1 and n.

Proof. Since $\frac{\partial F}{\partial y}(a,b) \neq 0$ for all solutions b of $F(a,b) = 0$, the polynomial $F(a,y)$ has exactly n roots $b_1, \ldots, b_n$. If $\varepsilon > 0$ is small and $y_i(x)$ the holomorphic function on $|x-a| < \varepsilon$ with $y_i(a) = b_i$ and $F\big(x, y_i(x)\big) \equiv 0$ (which exists by the theorem above), then the y_i have property (i) if ε is small enough, and property (ii) since the equation $F(x,\eta) = 0$ has at most n solutions.

Proposition 1. *Let $F \in \mathbb{C}[x,y]$ be irreducible. There are only finitely many $x \in \mathbb{C}$ such that the equations*

$$F(x,y) = 0 = \frac{\partial F}{\partial y}(x,y)$$

have a simultaneous solution $y \in \mathbb{C}$.

Proof. By the division algorithm, there are polynomials $b_i \in \mathbb{C}[x]$ $(i \geq 0)$ with $b_0 = a_0\big[F = a_0(x)y^n + \cdots + a_n(x)\big]$ and polynomials A_j, $Q_j \in \mathbb{C}[x,y]$ $(j \geq 1)$ such that

$$\begin{aligned} b_0^n F &= A_1 \frac{\partial F}{\partial y} + Q_1\,, \quad \deg_y Q_1 < \deg_y \frac{\partial F}{\partial y} = n-1 \\ b_1 \frac{\partial F}{\partial y} &= A_2 Q_1 + Q_2\,, \quad \deg_y Q_2 < \deg_y Q_1 \\ &\vdots \\ b_{k-1} Q_{k-2} &= A_k Q_{k-1} + Q_k\,, \quad \deg_y Q_k < \deg_y Q_{k-1}\,. \end{aligned}$$

We may suppose that $\deg_y Q_k = 0$, i.e. that $Q_k \in \mathbb{C}[x]$ (since we can otherwise continue the division process). We claim now that $Q_k(x) \not\equiv 0$. If, in fact, $Q_k \equiv 0$, then from the last of the above equations, any prime factor P of Q_{k-1} with $\deg_y P > 0$ would divide $b_{k-1}Q_{k-2}$, hence Q_{k-2} (since $b_{k-1} \in \mathbb{C}[x]$ and $\deg_y P > 0$). From the equation

$b_{k-2}Q_{k-3} = A_{k-1}Q_{k-2} + Q_{k-1}$, it would follow that P divides $b_{k-2}Q_{k-3}$ and hence Q_{k-3}. Repeating this argument, P would divide all the Q_j $(j \geq 1)$, hence also $\frac{\partial F}{\partial y}$ and F, contradicting the irreducibility of F. Thus $Q_k = Q_k(x) \in \mathbb{C}[x]$ is $\not\equiv 0$.

If now $a, b \in \mathbb{C}$ and $F(a,b) = 0 = \frac{\partial F}{\partial y}(a,b)$, we see from the above equations that $Q_1(a,b) = 0$, then that $Q_2(a,b) = 0, \ldots, Q_k(a,b) = Q_k(a) = 0$. Since $Q_k \not\equiv 0$, the set

$$\{x \in \mathbb{C} \mid \exists y \in \mathbb{C} \quad \text{with} \quad F(x,y) = 0 = \frac{\partial F}{\partial y}(x,y)\} \subset \{x \in \mathbb{C} \mid Q_k(x) = 0\}$$

is finite.

Before proceeding further, we insert some toplogical preliminaries. All topological spaces we consider will be Hausdorff.

Definition. A continuous map $p : X \to Y$, where X, Y are locally compact (Hausdorff) spaces, will be called *proper* if, for any compact set $K \subset Y$, the inverse image $p^{-1}(K)$ is compact in X

Lemma 2. *If X, Y are locally compact, a proper map $p : X \to Y$ is necessarily closed, i.e. takes closed sets in X to closed sets in Y.*

Proof. Let $A \subset X$ be closed, and $y_0 \in Y$. Let K be a compact neighbourhood of y_0 in Y. Then $p(A) \cap K = p(A \cap p^{-1}(K))$ is compact (since A is closed and $p^{-1}(K)$ is compact), hence closed in K.

Remark. A continuous map $p : X \to Y$ between locally compact spaces X, Y is proper, if and only if, for any locally compact topological space Z, the product

$$p \times \mathrm{id}_Z : X \times Z \longrightarrow Y \times Z \,, \quad (x,z) \longmapsto (p(x), z)$$

is closed. If X, Y have countable bases, this can be seen by using the following remark: if $\{x_1, \ldots, x_n, \ldots\}$ is a sequence of points in X, without limit points and such that $\{p(x_n)\}_{n \geq 1}$ converges in Y, then the image of the closed set $\{(x_n, \frac{1}{n}) \mid n \geq 1\}$ in $X \times \mathbb{R}$ is not closed in $Y \times \mathbb{R}$.

The property in this remark can be used to define proper mappings between spaces which are not locally compact.

Remark. Let $p : X \to Y$ be a proper map between locally compact spaces. Let $Z \subset Y$ be a locally compact space (with the induced topology). Then $p \mid p^{-1}(Z) : p^{-1}(Z) \to Z$ is again proper.

In fact, a compact subset of Z is a compact subset of Y.

Lemma 3. *Let $c_1, \ldots, c_n \in \mathbb{C}$. Let $w \in \mathbb{C}$ and suppose that $w^n + c_1 w^{n-1} + \cdots + c_n = 0$. Then*

$$|w| < 2 \max_\nu |c_\nu|^{1/\nu}$$

(unless $c_1 = \cdots = c_n = 0$).

Proof. Let $c = \max_\nu |c_\nu|^{1/\nu} > 0$. If $z = \frac{w}{c}$, we have $z^n + \frac{c_1}{c} z^{n-1} + \cdots + \frac{c_n}{c^n} = 0$, so that, since $|c_\nu| \leq c^\nu$,

$$|z|^n \leq |z|^{n-1} + \cdots + 1 \, .$$

If $|z| \geq 2$, we would have $1 \leq \frac{1}{|z|} + \cdots + \frac{1}{|z|^n} \leq \frac{1}{2} + \cdots + \frac{1}{2^n} < 1$, a contradiction. Thus $|z| < 2$, i.e. $|w| < 2c$.

Proposition 2. *Let $F \in \mathbb{C}[x,y], F(x,y) = a_0(x)y^n + \cdots + a_n(x), a_0 \not\equiv 0$. Let $V = \{(x,y) \in \mathbb{C}^2 \mid F(x,y) = 0\}$ and $S_0 = \{x \in \mathbb{C} \mid a_0(x) = 0\}$. Let $\pi : V \to \mathbb{C}$ be the projection $(x,y) \mapsto x$. Then $\pi \mid \pi^{-1}(\mathbb{C} - S_0) \to \mathbb{C} - S_0$ is proper.*

Proof. Let $K \subset \mathbb{C} - S_0$ be compact. Then there is $\delta > 0$ so that $|a_0(x)| \geq \delta$ and $|a_\nu(x)| \leq \frac{1}{\delta}$ for $x \in K$. If $(x,y) \in V$, $x \in \pi^{-1}(K)$, we have

$$y^n + \frac{a_1(x)}{a_0(x)} y^{n-1} + \cdots + \frac{a_n(x)}{a_0(x)} = 0 \, ,$$

so that, by (1.8), $|y| \leq 2 \max_\nu \delta^{-2/\nu}$. Thus $\pi^{-1}(K)$ is bounded. Since clearly $\pi^{-1}(K) = (K \times \mathbb{C}) \cap V$ is closed in $\mathbb{C}^2$, $\pi^{-1}(K)$ is compact.

Definition. Let X, Y be (Hausdorff) topological spaces and $p : X \to Y$, a continuous map. p is called a covering map if the following holds: $\forall y_0 \in Y$, there is an open neighbourhood V of y_0 such that $p^{-1}(V)$ is a disjoint union $\bigcup_{j \in J} U_j$ of open sets U_j with the property that $p \mid U_j$ is a homeomorphism onto V $\forall j \in J$. The triple (X, Y, p) is then called an (unramified) covering. We also say that X is a covering of Y.

An open set $V \subset Y$ with the property in the definition is said to be evenly covered by p.

It follows from the definition that the cardinality of $p^{-1}(y)$ is a locally constant function on Y. (With the notation in the definition, the cardinality of $p^{-1}(y)$ is that of $J \forall y \in V$.) Thus, if Y is connected, "the number of points" in $p^{-1}(y)$ is independent of $y \in Y$. The covering is said to be finte (infinite) if the cardinality of $p^{-1}(y)$ is finite (infinite). p is called an n sheeted covering if $p^{-1}(y)$ contains exactly n points for $y \in Y$.

If $p : X \to Y, p : X \to T$ are two coverings of Y, they are said to be isomorphic if there exists a homeomorphism $\varphi : X' \to X$ such that $p \circ \varphi = p'$.

Examples. 1) Let $\Delta = \{z \in \mathbb{C} \mid |z| < 1\}$ and $\Delta^* = \Delta - \{0\}$. Then, if $n \geq 1$, the map $p_n : \Delta^* \to \Delta^*$ given by $p_n(z) = z^n$ is an n-sheeted covering.

It is a standard fact in the theory of covering spaces that any connected n-sheeted covering of Δ^* is isomorphic to p_n.

2) $p : \mathbb{C} \to \mathbb{C}^*$, $p(z) = e^z$ is an infinite covering of $\mathbb{C}^*$.

3) Let X, Y be locally compact, let $p : X \to Y$ be a local homeomorphism (i.e. $\forall a \in X$, $\exists$ an open neighbourhood U of a such that $V = p(U)$ is open in Y and $p \mid U$ is a homeomorphism onto V).

Then, p is a finite covering if and only if it is proper.

Proof. If p is a finite covering, if $y_0 \in Y$ and V is an open neighbourhood of y_0 which is evenly covered by p, then $p \mid p^{-1}(V) \to V$ is clearly proper. It follows easily that p is proper.

Conversely, let p be a proper local homeomorphism, let $y_0 \in Y$ and let $p^{-1}(y_0) = \{x_1, \ldots, x_n\}$. Let U_j' be an open set with $x_j \in U_j'$ and such that $p \mid U_j$ is a homeomorphism onto the open set $V_j = p(U_j')$. Since p is proper and $X - \bigcup_1^n U_j'$ is closed in X, $E = p(X - \bigcup_1^n U_j')$ is closed in Y. Clearly, $y_0 \notin E$. Let $V = Y - E$. Then $p^{-1}(V) \subset U_1' \cup \cdots \cup U_n'$, and we have $V \subset V_1 \cap \cdots \cap V_n$. If we set $U_j = U_j' \cap p^{-1}(V)$, then $p^{-1}(V) = \bigcup_1^n U_j$ and $p \mid U_j$ is a homeomorphism onto V.

Let $F \in \mathbb{C}[x, y]$ be irreducible, $F(x, y) = a_0(x)y^n + \cdots + a_n(x)$. Let $S_0 = \{x \in \mathbb{C} \mid a_0(x) = 0\}$ and $S_1 = \{x \in \mathbb{C} \mid \exists y \in \mathbb{C} \text{ with } F(x, y) = 0 = \frac{\partial F}{\partial y}(x, y)\}$. Then, if $V = \{(x, y) \in \mathbb{C}^2 \mid F(x, y) = 0\}$ and $\pi : V \to \mathbb{C}$ the projection $(x, y) \mapsto x$, then

$$\pi \mid \pi^{-1}(\mathbb{C} - (S_0 \cup S_1)) \longrightarrow \mathbb{C} - (S_0 \cup S_1)$$

is a finite covering (of n sheets).

This follows from Proposition 2 the implicit function theorem.

Before proceeding to show how the set V can be modified over the points of $S_0 \cup S_1$ and the point at ∞ in $\mathbb{C}$ to define the algebraic function completely, we shall introduce the notion of a Riemann surface and some related topics.

2. Riemann Surfaces

Let X be a 2-dimensional manifold (i.e. X is a Hausdorff space and any point in X has a neighbourhood homeomorphic to an open set in $\mathbb{R}^2$).

Consider pairs (U, φ) where U is open in X and $\varphi : U \to \varphi(U) \subset \mathbb{C}$ is a homeomorphism onto an open set in $\mathbb{C}$.

Two such pairs (U_1, φ_1), (U_2, φ_2) are said to be (holomorphically) compatible if the map $\varphi_2 \circ \varphi_1^{-1} : \varphi_1(U_1 \cap U_2) \to \varphi_2(U_1, \cap U_2)$ is holomorphic; its inverse is also holomorphic by a standard result in complex analysis.

A complex structure on X is a family $\mathcal{S}$ of pairs $\{(U, \varphi)\}$ which are pairwise compatible and such that $\bigcup U = X$; there is then a unique maximal family of pairs with these two properties and containing $\mathcal{S}$; we shall usually assume that the complex structure is maximal. The elements (U, φ) of this (maximal) complex structure are called charts or coordinate neighbourhoods. In a coordinate neighbourhood, we usually identify U with $\varphi(U)$ and write z for φ as one does with the usual complex variable in $\mathbb{C}$.

A Riemann surface is a connected 2-dimensional manifold X with a complex structure $\mathcal{S}$. We shall also assume that X has a countable base of open sets, although a theorem of Radó asserts that this is automatic (for a proof, see e.g. [4]).

If $\Omega \subset X$ is open (X is a Riemann surface) and $f : \Omega \to \mathbb{C}$ is continuous, we say that f is holomorphic if for any chart (U, φ) of X, the function $f \circ \varphi^{-1} : \varphi(\Omega \cap U) \to \mathbb{C}$ is holomorphic.

If X, Y are Riemann surfaces, $f : X \to Y$ a continuous map, f is called holomorphic if, for any chart (V, ψ) of Y, the function $\psi \circ f : f^{-1}(V) \to \psi(V) \subset \mathbb{C}$ is holomorphic.

Non-constant holomorphic maps between Riemann surfaces are open. Also, a bijective holomorphic map $f : X \to Y$ has a holomorphic inverse $f^{-1} : Y \to X$. Such bijective holomorphic maps are called analytic isomorphisms (or biholomorphic maps).

Examples

1. *The complex projective line = Riemann sphere.* Let $\mathbb{P}^1$ be the one-point compactification $\mathbb{C} \cup \{\infty\}$ of $\mathbb{C}$. We set $U_1 = \mathbb{P}^1 - \{\infty\} = \mathbb{C}$, $\varphi_1 : U_1 \to \mathbb{C}$ being the identity;

$$U_2 = \mathbb{P}^1 - \{0\},\ \varphi_2(z) = \begin{cases} 1/z & \text{if } z \in \mathbb{C} - \{0\} = \mathbb{C}^* \\ 0 & \text{if } z = \infty\,. \end{cases}$$

The map $\varphi_2 \circ \varphi_1^{-1}$ is the map $z \mapsto 1/z$ of $\mathbb{C}^*$ into itself, so that these two charts define a complex structure on $\mathbb{P}^1$. This Riemann surface is called the projective line or the Riemann sphere.

2. *Tori*. Let $\tau \in \mathbb{C}$, $\mathrm{Im}(\tau) > 0$. Let $\Lambda = \{m+n\tau \mid m,n \in \mathbb{Z}\}$. Λ is an additive subgroup of $\mathbb{C}$. Consider the quotient group $X = \mathbb{C}/\Lambda$ and let $\pi : \mathbb{C} \to X$ be the canonical projection. With the quotient topology, X is a compact Hausdorff space, and $\mathbb{C} \to X$ is a local homeomorphism. [These statements are easy consequences of the following two remarks: if $a \in \mathbb{C}$, and we consider the set $U = \{a+\lambda+\mu\tau \mid \lambda, \mu \in \mathbb{R}, -\frac{1}{2} < \lambda, \mu < +\frac{1}{2}\}$, U is open and maps bijectively onto an open set in X; further X is the image of the compact set $\bar{U}$ (closure of U) for any $a \in \mathbb{C}$. π is actually a covering map.]

As charts, we use pairs (U, φ) obtained as follows: let V be any open set in $\mathbb{C}$ such that $\pi|V$ is a homeomorphism onto an open set U in X; set $\varphi = (\pi|V)^{-1} : U \to V \subset \mathbb{C}$. Two such charts (U_1, φ_1), (U_2, φ_2) are holomorphically compatible: we clearly have $\pi(\varphi_2 \circ \varphi_1^{-1}(z)) = \pi(z)$ for $z \in \varphi_1(U_1 \cap U_2)$ thus $\varphi_2 \circ \varphi_1^{-1}(z) - z \in \Lambda\ \forall z \in \varphi_1(U_1 \cap U_2)$, so must be constant on connected components (because $\varphi_2 \circ \varphi_1^{-1}$ is continuous and Λ is discrete).

The Riemann surfaces X constructed above are called tori or elliptic curves.

3. *Surfaces of "higher genus"*. Let g be an integer > 1, and let $0 < r < 1$. Let $\Delta = \{z \in \mathbb{C} \mid |z| < 1\}$. There is a unique bijective holomorphic (= biholomorphic) map $T : \Delta \to \Delta$ such that $T(r) = re^{3\pi i/2g}$ and $T(re^{\pi i/2g}) = re^{2\pi i/2g}$. Let $\sigma : \Delta \to \Delta$ be the rotation $z \mapsto ze^{2\pi i/4g}$.

For $k \in \mathbb{Z}$, we set

$$A_k = \sigma^{4k} T \sigma^{-4k}\ , \quad B_k = \sigma^{4k+1} T \sigma^{-4k-1}\ ,$$

and denote by Γ the group of biholomorphic maps of Δ generated by $A_k, B_k (\forall k \in \mathbb{Z})$.

A special case of a theorem enunciated by Poincaré (for the theorem and its proof, see the elegant article by G. de Rham: *Sur les polygônes générateurs de groupes Fuchsiens*, L'Enseignement Mathématique, 1971, pp. 47–61) implies that there exists an r, $0 < r < 1$, such that Γ acts freely (without fixed points) and discontinuously on Δ, and the quotient Δ/Γ is compact. One sees that the canonical projection $\pi : \Delta \to \Delta/\Gamma$ is a covering map, and obtains a complex structure on Δ/Γ for which the map π is holomorphic as in the case of tori.

4. Let Y be a Riemann surface, X a connected 2-dimensional manifold and $p : X \to Y$ a local homeomorphism. There is a unique complex structure on X for which the map p is holomorphic, obtained as follows: Let U be an open set in X such that $p|U$ is a homeomorphism onto an open set V in Y such that $V \subset V_j$ for some j, where $\{(V_j, \psi_j)_{j \in J}\}$ is the given complex structure on Y. Let $\varphi_U : U \to \mathbb{C}$ be the map $\varphi_U = \psi_j \circ p$. It is easily checked that two such pairs (U, φ_U), $(U', \varphi_{U'})$ are holomorphically compatible, so that one obtains a complex structure on X for which p is holomorphic.

The uniqueness is a consequence of the following remark: let $U \subset X$ be open and $p|U$, a homeomorphism onto $V \subset Y$. Then, if p is holomorphic, the map $(p|U)^{-1} : V \to U$ is again holomorphic.

Consider now a Riemann surface X and a holomorphic map $p : X \to \mathbb{C}$ which is also a local homeomorphism. We consider $\mathbb{C}$ as the complement of $\infty \in \mathbb{P}^1$, and p as a local homeomorphism $X \to \mathbb{P}^1$.

We shall define boundary points of X. Let $\{x_\nu\}_{\nu \geq 1}$ be a sequence of points in X with the following properties:

1) $\{x_\nu\}$ is discrete (i.e. has no limit points in X);

2) $\{p(x_\nu)\}$ converges to a point $a \in \mathbb{P}^1$;

3) Let $D_\varepsilon = \{z \in \mathbb{C} \mid |z - a| < \varepsilon\}$ if $a \in \mathbb{C}$, and let $D_\varepsilon = \{z \in \mathbb{C} \mid |z| > \frac{1}{\varepsilon}\} \cup \{\infty\}$ if $a = \infty$. Then, for all sufficiently small $\varepsilon > 0$, all but finitely many of the $\{x_\nu\}$ lie in the same connected component of $p^{-1}(D_\varepsilon)$.

Two such sequences $\{x_\nu\}, \{y_\nu\}$ are called equivalent if the sequence

$$z_\nu = \begin{cases} x_{(\nu+1)/2} & \text{for } \nu \text{ odd} \\ y_{\nu/2} & \text{for } \nu \text{ even} \end{cases}$$

again has the three properties above [i.e. $\lim p(x_\nu) = \lim p(y_\nu) = a$ say, and the connected components of $p^{-1}(D_\varepsilon)$ containing all but finitely many of the x_ν, y_ν respectively are the same].

A *boundary point* of X (relative to the map p) is then an equivalence class of sequences $\{x_\nu\}_{\nu \geq 1}$ with the three properties given above. Set $\tilde{X} = X \cup \{\text{boundary points of } X\}$.

Let P be a boundary point of X, defined by a sequence $\{x_\nu\}_{\nu \geq 1}$. We define neighbourhoods of P in $\tilde{X}$ as follows. Let $\varepsilon > 0$ be small and $D_\varepsilon = \{z \mid |z - a| < \varepsilon\}$ ($a \in \mathbb{C}$) or $D_\varepsilon = \{z \mid |z| > \frac{1}{\varepsilon}\} \cup \{\infty\}$ ($a = \infty$), where $a = \lim p(x_\nu)$. Let Ω_ε be the connected component of $p^{-1}(D_\varepsilon)$ containing all but finitely many of the x_ν, and let $\tilde{\Omega}_\varepsilon$ be the union of Ω_ε with those boundary points Q with the following property: if $\{y_\nu\}_{\nu \geq 1}$ defines Q, then $\{\nu \mid y_\nu \notin \Omega_\varepsilon\}$ is finite (this is independent of the sequence $\{y_\nu\}$ defining Q). The $\tilde{\Omega}_\varepsilon$ ($\varepsilon > 0$ small) form a fundamental system of neighbourhood of $P \in \tilde{X} - X$.

This topology is Hausdorff: if P, Q are boundary points defined by $\{x_\nu\}, \{y_\nu\}$ respectively, and $P \neq Q$, then, by the definition of the equivalence relation, there is $\varepsilon > 0$ such that the components $\Omega_{\varepsilon,1}\Omega_{\varepsilon,2}$ of $p^{-1}(D_\varepsilon)$ containing all but finitely many of the x_ν, y_ν respectively are distinct, and $\tilde{\Omega}_{\varepsilon,1} \cap \tilde{\Omega}_{\varepsilon,2} = \emptyset$. Moreover, p clearly extends to a continuous map $\tilde{p} : \tilde{X} \to \mathbb{P}^1 : \tilde{p}(P) = a = \lim p(x_\nu)$.

A boundary point P of X is said to the *algebraic* if the following holds: let D_ε be a small disc around $a = \tilde{p}(P)$ and let Ω be the connected component of $p^{-1}(D_\varepsilon)$ containing all but finitely many points of a sequence defining P; then $p(\Omega) \subset D_\varepsilon - \{a\}$ and the map $p : \Omega \to D_\varepsilon - \{a\}$ is a finite covering.

If we set $\Delta_R = \{z \in \mathbb{C} \mid |z| < R\}$ and $\Delta_R^* = \Delta_R - \{0\}$, then there is $n \geq 1$ such that the map $p : \Omega \to D_\varepsilon - \{a\}$ is isomorphic to the map $p_n : \Delta^*_{\varepsilon^{1/n}} \to D_\varepsilon - \{a\}$ given by $p_n(z) = a + z^n$ if $a \in \mathbb{C}$, $p_n(z) = z^{-n}$ if $a = \infty$ (see Example 1 after Definition (1.10)).

In this case, $\tilde{\Omega} = \Omega \cup \{P\}$ is a neighbourhood of P in $\tilde{X}$ containing no other boundary points of X. Since $p|\Omega \to D_\varepsilon - \{a\}$ is isomorphic to the map p_n defined above, there is a homeomorphism $\varphi : \tilde{\Omega} \to \Delta_{\varepsilon^{1/n}}$ with $\varphi(P) = 0$ and $p \circ \varphi^{-1} = p_n$ on $\Delta^*_{\varepsilon^{1/n}}$. Clearly, $\varphi|\Omega$ is holomorphic.

Set $\hat{X} = X \cup \{\text{algebraic boundary points of } X\}$. We can extend the complex structure on X to one on $\hat{X}$ by taking as a chart containing an algebraic boundary point $P \in \hat{X} - X$ the pair $(\tilde{\Omega}, \varphi)$ constructed above. Let $\hat{p} = \tilde{p}|\hat{X}$. The pair $(\hat{X}, \hat{p})$ will be called the (algebraic) completion of (X, p). The map $\hat{p} : \hat{X} \to \mathbb{P}^1$ is holomorphic, but does not have to be a local homeomorphism. With the notation above, if P is an algebraic boundary point and Ω is an n-sheeted covering of $D_\varepsilon - \{a\}$ with $n > 1$, then $\hat{p}$ is not a local homeomorphism at P.

This construction can be used to obtain the Riemann surface of a holomorphic function as conceived by Riemann. To do this, we first introduce the sheaf of germs of holomorphic functions on a Riemann surface.

3. The Sheaf of Germs of Holomorphic Functions

Let X be a Riemann surface, and let $a \in X$. We consider pairs (U, f), where U is an open neighbourhood of a and f is a holomorphic function on U. Two such pairs (U, f) and (V, g) are said to be equivalent, and define the same germ of holomorphic function at a, if there exists an open neighbourhood W of a, $W \subset U \cap V$, such that $f|W = g|W$. An equivalence class is called a germ of holomorphic function at a; the class of a pair (U, f) is called the germ of f at a and denoted by $\underline{f}_a$. The value at a of $\underline{f}_a$ is defined by $\underline{f}_a(a) = f(a)$ for any pair (U, f) defining $\underline{f}_a$.

If we choose a chart (U, φ) with $a \in U$ and $\varphi(a) = 0$, and identify functions on subsets of U with functions on subsets of $\varphi(U)$, then we can also speak of the values of the derivatives of a germ $\underline{f}_a$ at a:

$$\underline{f}_a^{(k)}(a) = \left(\frac{d}{dz}\right)^k f \circ \varphi^{-1} \Big|_{z=0} \quad \text{if} \quad (V, f) \quad \text{is a pair}$$

defining $\underline{f}_a$.

Let $\mathcal{O}_a$ be the set of all germs at a. $\mathcal{O}_a$ is clearly a ring, even a $\mathbb{C}$-algebra. The set m_a of germs $\underline{f}_a$ with $\underline{f}_a(a) = 0$ is an ideal; the complement $\mathcal{O}_a - m_a$ consists of the units of $\mathcal{O}_a$ ($\underline{f}_a$ has an inverse in $\mathcal{O}_a \iff \underline{f}_a(a) \neq 0$) so that m_a is the unique maximal ideal in $\mathcal{O}_a$.

If we choose a chart as above, then the map $\underline{f}_a \longmapsto \sum_{n=0}^{\infty} \frac{1}{n!} \underline{f}_a^{(n)}(a) z^n$ is an isomorphism (as $\mathbb{C}$-algebras) of $\mathcal{O}_a$ with the ring $\mathbb{C}\{z\}$ of power series with a non-zero radius of convergence.

Let $\mathcal{O}_X = \bigcup_{a \in X} \mathcal{O}_a$ (disjoint union). We will sometimes write simply $\mathcal{O}$ for $\mathcal{O}_X$. We define a map $p : \mathcal{O}_X \to X$ by $p(\underline{f}) = a$ if $\underline{f} \in \mathcal{O}_X$.

Let now $\underline{f}_a \in \mathcal{O}_X$, and (U, f) a pair defining the germ $\underline{f}_a$. We set $N(U, f) = \{\underline{f}_x \mid x \in U\}$, the set of all germs defined by f at the different points of U. We define a topology on $\mathcal{O}_X$ by the condition that the sets $\{N(U, f)\}$ form a fundamental system of neighbourhoods of $\underline{f}_a$ when (U, f) runs over all pairs defining $\underline{f}_a$.

Lemma. *With this topology, $\mathcal{O}_X$ is Hausdorff and the map $p : \mathcal{O}_X \to X$ is a local homeomorphism.*

Proof. Let $\underline{f}_a, \underline{g}_b \in \mathcal{O}_X$, and suppose that $\underline{f}_a \neq \underline{g}_b$. We must show that they can be separated by open neighbourhoods, and consider two cases.

1) $a \neq b$. Let (U, f), (V, g) be pairs defining $\underline{f}_a$, $\underline{g}_b$ respectively ($a \in U, b \in V$). We can find such pairs with $U \cap V = \emptyset$, in which case $N(U, f) \cap N(V, g) = \emptyset$.

2) $a = b$. Let U be a connected open set, $a \in U$, and f, g, holomorphic functions on U so that the pairs (U, f), (U, g) define $\underline{f}_a$, $\underline{g}_a$ respectively. We claim that $N(U, f) \cap N(U, g) = \emptyset$ in fact, if $\underline{h}_x$ $(x \in U)$ is a germ in the intersection, then both f and g induce the germ $\underline{h}_x$ at x, hence coincide in some neighbourhood of x. Since U is connected, the principle of analytic continuation implies that $f \equiv g$, so that $\underline{f}_a = \underline{g}_a$, a contradiction.

Thus, $\mathcal{O}_X$ is Hausdorff.

Moreover, if U is open in X and f holomorphic on U, then $p(N(U, f)) = U$ and $p|N(U, f)$ is injective, having the inverse $x \mapsto \underline{f}_x$ (= germ of f at x).

The properties of p stated in the lemma follow from this.

Remark. If X is given with a countable base for its open sets, it can be proved directly that any connected component of $\mathcal{O}_X$ has a countable base. This is a consequence of the so-called Poincaré–Volterra theorem (for a statement and proof of which one may consult [7]).

We can now construct the "Riemann surface of an analytic function".

Let $X = \mathbb{C}$ and consider $\mathcal{O}_{\mathbb{C}}$. Let M be a connected component of $\mathcal{O}_{\mathbb{C}}$, and $p : M \to \mathbb{C} \subset \mathbb{P}^1$ the restriction to M of the map $\underline{f}_a \mapsto a$ constructed earlier; $p : M \to \mathbb{C}$ is a local homeomorphism, so there is a unique structure of a Riemann surface on M for which p is holomorphic; p is then a local analytic isomorphism, i.e any $a \in M$ has an open neighbourhood U such that $p|U$ is an analytic isomorphism of U onto the open set $p(U)$.

We define a holomorphic function h on M by $h(\underline{f}_a) = \underline{f}_a(a)$ [if (U, f) is a pair defining $\underline{f}_a$ with U connected, then $N(U, f) \subset M$ and $h(\underline{f}_x) = f(x)$ $\forall x \in U$, so that h is holomorphic]. Intuitively, this "universal" function h describes all the germs that can be obtained by "analytic continuation" of a fixed germ $\underline{f}_a \in M$.

Let now $\hat{M} = M \cup \{\text{algebraic boundary points of } M\}$ and $\hat{p} : \hat{M} \to \mathbb{P}^1$ be the holomorphic map extending $p : M \to \mathbb{C}$ we constructed before.

Let U be a connected open set in $\mathbb{C}$ and f, a holomorphic function on U. We assume that (U, f) defines a germ $\underline{f}_a \in M$ for some $a \in U$ (and hence for all $a \in U$, since U is connected).

The set $E = \hat{M} - M$ is a discrete set in $\hat{M}$, and we have our universal function h on M. Let X_f be the union of M with those points $P \in E$ which are not essential singularities of h (i.e. points where h has either a holomorphic extension or a pole). Thus, there is a meromorphic function h_f on X_f with $h_f|M = h$. Let $p_f : X_f \to \mathbb{P}^1$ be the restriction of $\hat{p}$.

The triple (X_f, p_f, h_f) *is the Riemann surface of the function* f *on* U.

If $P \in X_f - M$, and $p_f(P) = a$, then near P, the map p_f is equivalent to the map $z \mapsto a + z^n$ $(n \geq 1)$ or $z \mapsto z^{-n}$. Thus, if z is a local coordinate at P on X_f and w

denotes a local coordinate at a on $\mathbb{P}^1$, the local description of the maps p_f, h_f can be written:

$$\begin{cases} w = z^n \\ h_f = \sum_{-k}^{\infty} a_\nu w^{\nu/n} \quad (= \sum_{-k}^{\infty} a_\nu z^\nu) \,. \end{cases}$$

4. The Riemann Surface of an Algebraic Function

Let $F(x,y) = a_0(x)y^n + a_1(x)y^{n-1} + \cdots + a_n(x) \in \mathbb{C}[x,y]$ be an irreducible polynomial, let $V = \{(x,y) \in \mathbb{C}^2 \mid F(x,y) = 0\}$. Let $S_0 = \{x \in \mathbb{C} \mid a_0(x) = 0\}$, $S_1 = \{x \in \mathbb{C} \mid \exists y \in \mathbb{C}$ with $F(x,y) = 0 = \frac{\partial F}{\partial y}(x,y)\}$ and let $S = S_0 \cup S_1 \cup \{\infty\} \subset \mathbb{P}^1$. Let $\pi : V \to \mathbb{C}$ be the projection $(x,y) \mapsto x$, $V' = V - \pi^{-1}(S) = V - \pi^{-1}(S_0 \cup S_1)$ and $\pi' = \pi|V'$. We have seen that if D_ε is a small disc around $a \in S$ $\left(D_\varepsilon = \{|z| > \frac{1}{\varepsilon}\} \cup \{\infty\}\text{ if } a = \infty\right)$, then $\pi'|\pi^{-1}(D_\varepsilon - \{a\}) \to D_\varepsilon - \{a\}$ is a finite covering (of n-sheets). In particular, $\pi^{-1}\left(D_\varepsilon - \{a\}\right)$ has only finitely many connected components. Moreover, if W is a connected component of V', then $\pi'|W$ is again a covering, and so maps W onto $\mathbb{P}^1 - S$. Hence V' has only finitely many connected components. (We shall see below that it is, in fact, connected.) Let $W_1, \ldots, W_r$ be the components of V'. Then $\pi_j = \pi|W_j \to \mathbb{P}^1 - S$ is a finite covering, hence every boundary point of W_j is algebraic. Let $\hat{\pi}_j : \hat{W}_j \to \mathbb{P}^1$ be the algebraic completion of $\pi_j : W_j \to \mathbb{P}^1 - S$. If $P \in \hat{W}_j - W_j$ and $a = \hat{\pi}_j(P)$, there is a neighbourhood U of P and $\varepsilon > 0$ such that $\hat{\pi}_j|U \to D_\varepsilon$ is isomorphic to the map $z \mapsto a + z^m$ (or $z \mapsto z^{-m}$) for some $m > 0$, so that, in particular, $\hat{\pi}_j|U \to D_\varepsilon$ is proper. It follows that for any $a \in S$, there is $\varepsilon > 0$ so that $\hat{\pi}_j|\hat{\pi}_j^{-1}(D_\varepsilon) \to D_\varepsilon$ is proper. Since $\hat{\pi}_j|W_j \to \mathbb{P}^1 - S$ is proper, $\hat{\pi}_j : \hat{W}_j \to \mathbb{P}^1$ is proper, so that $\hat{W}_j$ is compact.

Let $p_2 : V \to \mathbb{C}$ be the second projection $(x,y) \mapsto y$. Then the function $\eta = p_2|V'$ is holomorphic on V', so that $\eta_j = \eta|W_j$ is holomorphic.

We claim that η_j extends to a meromorphic function on $\hat{W}_j$. To see this, let $a \in S$. Let $P \in \hat{W}_j$, $\hat{\pi}_j(P) = a$, and choose local coordinates z at P and w at a so that $\hat{\pi}_j$ becomes the map $z \mapsto z^m = w$. If U is a small neighbourhood of P, by the definition of V and η, we have, if $z \neq 0$,

$$\eta_j^n + \frac{a_1(w)}{a_0(w)}\eta_j^{n-1}(z) + \cdots + \frac{a_n(w)}{a_0(w)} = 0\,, \quad w = \hat{\pi}_j(z)\,.$$

Moreover, the a_ν/a_0 are meromorphic at $w = 0$, so that there exist constants $C > 0$, $N > 0$ so that $\left|\frac{a_\nu(w)}{a_0(w)}\right| \leq \frac{C}{|w|^N}$ near $w = 0$. By Lemma 3, we have $|\eta_j(z)| \leq 2\max_\nu \frac{C^{1/\nu}}{|w|^{N/\nu}} \leq \frac{C_1}{|z|^k}$ for some constants C_1, k. Hence η_j has a meromorphic extension to $\hat{W}_j$.

We now claim that V' is connected. If this were not the case, $\pi_1 : W_1 \to \mathbb{P}^1 - S$ is an r-sheeted covering with $1 \leq r < n$.

For $x \in \mathbb{P}^1 - S$, let $b_\nu(x)$ $(\nu = 1, \ldots, r)$ be the νth-elementary symmetric function of $y_1, \ldots, y_r$, where the y_j are the values taken by η_1 at the points of $\pi_1^{-1}(x)$; by the definition of η_1, the y_j are values of the second projection p_2 at points $(x,y) \in V$, so that $F(x,y_j) = 0$, $j = 1,\ldots,r$.

We claim that b_ν extend to meromorphic functions on $\mathbb{P}^1$. In fact, since the y_j are values of the function η_1 which is meromorphic on $\hat{W}_1$, in the neighbourhood of $a \in S$ we have an estimate of the form

$$\begin{aligned} |b_\nu(x)| = \left| \sum_{i_1,\ldots,i_\nu} \eta_1(P_{i_1})\ldots\eta_\nu(P_{i_\nu}) \right| &\le C\,|z|^{-\ell} \\ &\le C_1\,|x-a|^{-\ell'} \text{ (resp. } C_1\,|x|^{-\ell'} \text{ if } a=\infty) \\ \left[(P_1,\ldots,P_r) = \pi_1^{-1}(x)\right] . \end{aligned}$$

Thus, the b_ν are meromorphic on $\mathbb{P}^1$, and are therefore rational functions of x.

Let $G(x,y) = y^r + b_1(x)y^{r-1} + \ldots + b_r(x)$. Then, if $x \in \mathbb{P}^1 - S$, the roots of $G(x,y)$ (viz $y_1,\ldots,y_r$), are also roots of $F(x,y)$. Hence G divides F in $\mathbb{C}(x)[y]$, and, since $\deg_y G \ge 1$, F is not irreducible (Gauss' lemma).

Thus V' is connected, and $\hat{W}$, the algebraic completion of V', is a compact Riemann surface. $\hat{W}$ carries a meromorphic function η, and if $\hat{\pi} : \hat{W} \to \mathbb{P}^1$ is the extension of $\pi' : V' \to \mathbb{P}^1 - S$, we have

$$F\big(\hat{\pi}(P), \eta(P)\big) \equiv 0 \quad \text{on} \quad \hat{W} .$$

This construction is, of course, a special case of the construction of the Riemann surface of a holomorphic germ, once one has proved the connectedness of V'. This statement is equivalent to the following:

Let $a \in \mathbb{P}^1 - S$, and let $y_1,\ldots,y_n$ be the germs at a satisfying the equation $F\big(x, y_j(x)\big) = 0$. Then, for any j, there is a closed curve γ in $\mathbb{P}^1 - S$ starting at a such that analytic continuation of y_1 along γ leads to y_j.

5. Sheaves

Let X be a topological space. A *presheaf of abelian groups on* X consists of the following data.

1) An assignment $U \mapsto \mathcal{F}(U)$ of an abelian group $\mathcal{F}(U)$ to each open set $U \subset X\,(\mathcal{F}(\emptyset) = \{0\} =$ abelian group with just one element)

2) A family $(\rho_V^U)_{V \subset U}$ of group homomorphisms $\rho_V^U : \mathcal{F}(U) \to \mathcal{F}(V)$ whenever U, V are open sets with $V \subset U$ (called restriction maps) having the following properties:

   (a) $\rho_U^U =$ identity on $\mathcal{F}(U)$ $\forall$ open U;

   (b) if $W \subset V \subset U$ are three open sets, then

$$\rho_W^U = \rho_W^V \circ \rho_V^U .$$

If the groups $\mathcal{F}(U)$ have additional structure (rings, vector spaces, $\mathbb{C}$-algebras,...) we shall speak of presheaves of (rings, vector spaces, $\mathbb{C}$-algebras,...) if the restriction maps respect this additional structure.

Example. Let X be a Riemann surface, and, for $U \subset X$ open, let $\mathcal{O}(U)$ denote the $\mathbb{C}$-algebra of functions holomorphic on U. If $V \subset U$, the map $\rho_V^U : \mathcal{O}(U) \to \mathcal{O}(V)$ will be restriction: $f \mapsto f|V$.

A presheaf $(\mathcal{F}(U), \rho_V^U)$ will be called a *sheaf* if the following two conditions are satisfied. Let $U \subset X$ be open, $U = \bigcup_{i \in I} U_i$ where the $\{U_i\}$ are again open

(i) If $f, g \in \mathcal{F}(U)$ and $\rho_{U_i}^U(f) = \rho_{U_i}^U(g)$ $\forall i$, then $f = g$.

(ii) Given, for each i, an element $f_i \in \mathcal{F}(U_i)$, if $\rho_{U_i \cap U_j}^{U_i}(f_i) = \rho_{U_i \cap U_j}^{U_j}(f_j)$ $\forall i, j \in I$, then $\exists f \in \mathcal{F}(U)$ with $\rho_{U_i}^U(f) = f_i \forall i$.

We shall usually denote $\rho_V^U(f)$ by $f|V$ (as if we were actually dealing with restriction of a mapping).

We can associate a sheaf to any presheaf (by the construction used to define germs of holomorphic functions). Let $\mathcal{F} = (\mathcal{F}(U), \rho_V^U)$ be a presheaf on X, and let $a \in X$. On the disjoint union $\coprod_{a \in U} \mathcal{F}(U)$ we introduce an equivalence relation as follows: $f \in \mathcal{F}(U)$, $g \in \mathcal{F}(V)$ are equivalent if $\exists$ an open set W, $a \in W \subset U \cap V$ such that $f|W = g|W$. The set $\mathcal{F}_a$ of equivalence classes is called the *stalk* of the presheaf $\mathcal{F}$ at a. [It is also the direct limit of the directed system $(\mathcal{F}(U), \rho_V^U)$.] We can introduce a topology on $|\mathcal{F}| = \coprod_{a \in X} \mathcal{F}_a$ (disjoint union) by taking as a fundamental system of neighbourhoods of $\underline{f}_a \in \mathcal{F}_a$ the following sets: let $f \in \mathcal{F}(U)$, $a \in U$, be a representative of the equivalence class $\underline{f}_a$, and let $N(U,f) = \{\underline{f}_x | x \in U\}$, where $\underline{f}_x$ is the equivalence class in $\mathcal{F}_x$ defined by (U,f). $|\mathcal{F}|$ does not have to be Hausdorff, but the projection map $p : |\mathcal{F}| \to X$, $p(f) = a$ if $f \in \mathcal{F}_a$ is local homeomorphism. If we set $|\mathcal{F}|(U) = \{$set of sections of $|\mathcal{F}|$ over U, i.e. the set of continuous maps $s : U \to |\mathcal{F}|$ such that $p \circ s =$ identity$\}$, and let $r_V^U(s)$ be the restriction of the map $s : U \to |\mathcal{F}|$ to $V \subset U$, then $\left(|\mathcal{F}|(U), r_V^U\right)$ is a *sheaf*, the sheaf associated to the presheaf $\mathcal{F}$.

We now define morphisms between presheaves. Let $\mathcal{F} = \left(\mathcal{F}(U), \rho_V^U\right)$ and $\mathcal{G} = \left(\mathcal{G}(U), r_V^U\right)$ be presheaves on X. A morphism $\alpha : \mathcal{F} \to \mathcal{G}$ is the assignment, to each U open $\subset X$, of a morphism $\alpha_U : \mathcal{F}(U) \to \mathcal{G}(U)$ such that, if $V \subset U$, the diagram

$$\begin{array}{ccc} \mathcal{F}(U) & \xrightarrow{\alpha_U} & \mathcal{G}(U) \\ \Big\downarrow \rho_V^U & & \Big\downarrow r_V^U \\ \mathcal{F}(V) & \xrightarrow{\alpha_V} & \mathcal{G}(V) \end{array}$$

commutes. If α_U is an isomorphism for all U, then $\alpha : \mathcal{F} \to \mathcal{G}$ is called an isomorphism.

If $\alpha : \mathcal{F} \to \mathcal{G}$ is a morphism of presheaves, we define the kernel, $\ker \alpha$, of α to be the presheaf

$$\left\{\ker(\alpha_U), \rho_V^U \mid \ker(\alpha_U)\right\} .$$

If $\mathcal{F}$ and $\mathcal{G}$ are sheaves, so is $\ker(\alpha)$.

If we define the image $\operatorname{im}(\alpha)$ to be the presheaf

$$\left\{\operatorname{im}(\alpha_U), r_V^U \mid \operatorname{Im}(\alpha_U)\right\} ,$$

then, even if $\mathcal{F}$ and $\mathcal{G}$ are sheaves, $\operatorname{Im}(\alpha)$ does not have to be a sheaf.

Example. Let $X = \mathbb{C}^* = \mathbb{C} - \{0\}$, let $\mathcal{O}(U) = \{$set of functions holomorphic on $U\}$, $\mathcal{O}^*(U) = \{$ set of functions holomorphic and nowhere zero on $U\}$. If $\exp : \mathcal{O} \to \mathcal{O}^*$ is the morphism defined by $\exp_U : f \mapsto \exp(2\pi i f)$ $\left(f \in \mathcal{O}(U)\right)$, then im(exp) fails to satisfy the second condition in the definition of sheaf. Namely, if $U_1 = \mathbb{C} - \{x \in \mathbb{R} | x \le 0\}$, $U_2 = \mathbb{C} - \{x \in \mathbb{R} | x \ge 0\}$, and we set $f_1(z) \equiv z$ on U_1, $f_2(z) \equiv z$ on U_2, then $f_i \in \operatorname{im}(\exp_{U_i})$ since U_1, U_2 are simply connected, but there is no $f \in \operatorname{im}(\exp_{U_1 \cup U_2})$ with $f|U_i = f_i$ $(i = 1, 2)$ (the function z has no single valued logarithm on $\mathbb{C}^* = U_1 \cup U_2$).

Remark. If $\mathcal{F}$ is a presheaf and we construct $|\mathcal{F}|$ the sheaf associated to $\mathcal{F}$, we have a morphism $\alpha : \mathcal{F} \to |\mathcal{F}|$ defined as follows: for $f \in \mathcal{F}(U)$, $\alpha_U(f)$ is the section of $|\mathcal{F}|$ over U defined by $a \mapsto \underline{f}_a =$ element of $\mathcal{F}_a$ induced by (U, f), $a \in U$. It can be checked directly that if $\mathcal{F}$ is a *sheaf*, then α is an isomorphism.

If we start with the sheaf $U \mapsto \mathcal{O}(U) = \{$space of functions holomorphic on $U\}$ on a Riemann surface X, then the space $|\mathcal{O}|$ is simply the "sheaf of germs of holomorphic functions on X" as defined in §3. Since the natural map from $\mathcal{O}$ to $|\mathcal{O}|$ is an isomorphism, we shall not distinguish between the two.

This is often called also the structure sheaf $\mathcal{O}_X$ of X.

Definition. If $\alpha : \mathcal{F} \to \mathcal{G}$ is a morphism between the sheaves $\mathcal{F}$ and $\mathcal{G}$, we shall denote by $\mathrm{Im}(\alpha)$ the *sheaf* associated to the presheaf $\{\mathrm{im}(\alpha_U), r_V^U \mid \mathrm{im}(\alpha_U)\}$.

Given morphisms $\alpha : \mathcal{E} \to \mathcal{F}$ and $\beta : \mathcal{F} \to \mathcal{G}$ between sheaves $\mathcal{E}, \mathcal{F}, \mathcal{G}$ on X, we say that the sequence

$$\mathcal{E} \xrightarrow{\alpha} \mathcal{F} \xrightarrow{\beta} \mathcal{G}$$

is exact (at $\mathcal{F}$) if the sheaves $\ker(\beta)$ and $\mathrm{Im}(\alpha)$ (in the sense just defined) are equal. This amounts to saying the following:

(a) $\beta_U \circ \alpha_U = 0\ \forall U$ and **(b)** if $f \in \mathcal{F}(U)$ and $\beta_U(f) = 0$, then, there exists an open covering $\{U_i\}_{i\in I}$ of U such that $f|U_i \in \mathrm{im}(\alpha_{U_i})\ \forall i \in I$.

Let now X be a topological space and $\mathcal{F}$ a sheaf of abelian groups on X. Let $\mathcal{U} = \{U_i\}_{i\in I}$ be an open covering of X. Then, for $q \geq 0$, we define the *group of q-cochains of $\mathcal{F}$ (relative to $\mathcal{U}$)* by

$$C^q(\mathcal{U}, \mathcal{F}) = \prod_{(i_0,\ldots,i_q)\in I^{q+1}} \mathcal{F}(U_{i_0} \cap \cdots \cap U_{i_q}) .$$

We define the coboundary $\delta : C^0(\mathcal{U}, \mathcal{F}) \to C^1(\mathcal{U}, \mathcal{F})$ by $\delta\big((f_i)_{i\in I}\big) = (c_{ij})_{i,j\in I}$, where $c_{ij} = f_i|U_i \cap U_j - f_j|U_i \cap U_j$. We also set

$$Z^1(\mathcal{U}, \mathcal{F}) = \{(c_{ij}) \in C^1(\mathcal{U}, \mathcal{F}) \mid c_{ij} + c_{jk} = c_{ik} \quad \text{on} \quad U_i \cap U_j \cap U_k\ \forall i, j, k \in I\}$$

(strictly speaking, the condition is that

$$c_{ij}\,|U_{ijk} + c_{jk}|\,U_{ijk} = c_{ik}|U_{ijk}\ , \quad U_{ijk} = U_i \cap U_j \cap U_k\ .)$$

Finally, let $B^1(\mathcal{U}, \mathcal{F}) = \mathrm{Image}(\delta : C^0(\mathcal{U}, \mathcal{F}) \to C^1(\mathcal{U}, \mathcal{F}))$; we have $B^1(\mathcal{U}, \mathcal{F}) \subset Z^1(\mathcal{U}, \mathcal{F})$.

We call the quotient group

$$H^1(\mathcal{U}, \mathcal{F}) = Z^1(\mathcal{U}, \mathcal{F})/B^1(\mathcal{U}, \mathcal{F})$$

the first cohomology group of $\mathcal{F}$ relative to $\mathcal{U}$.

We also set

$$H^0(\mathcal{U}, \mathcal{F}) = \{(f_i)_{i\in I} \in C^0(\mathcal{U}, \mathcal{F}) \mid \delta(f_i)_{i\in I} = 0\}\ ;$$

by the sheaf axioms, the map $\mathcal{F}(X) \to C^0(\mathcal{U}, \mathcal{F})$ defined by $f \mapsto (f|U_i)_{i\in I}$ induces an isomorphism of $\mathcal{F}(X)$ onto $H^0(\mathcal{U}, \mathcal{F})$ for any open covering $\mathcal{U}$. Elements of $\mathcal{F}(X) = H^0(\mathcal{U}, \mathcal{F})$ are also called (global) sections of $\mathcal{F}$.

Let $\mathcal{V} = (V_\alpha)_{\alpha\in A}$ be a refinement of $\mathcal{U}$; there is thus a refinement map $\tau : A \to I$ such that $V_\alpha \subset U_{\tau(\alpha)}$ $\forall \alpha \in A$ ($\mathcal{V}$ is also an open covering of X). τ induces a map

$$\tau^* : H^1(\mathcal{U},\mathcal{F}) \longrightarrow H^1(\mathcal{V},\mathcal{F})$$

as follows. If $\xi = (c_{ij})_{i,j\in I} \in Z^1(\mathcal{U},\mathcal{F})$, we define $\tau^*(\xi) = (\gamma_{\alpha\beta})_{\alpha\beta}$ by $\gamma_{\alpha\beta} = c_{\tau(\alpha)\tau(\beta)}|V_\alpha \cap V_\beta$. Clearly $\tau^*\big(B^1(\mathcal{U},\mathcal{F})\big) \subset B^1(\mathcal{V},\mathcal{F})$ so that it induces a map (denoted again τ^*) of $H^1(\mathcal{U},\mathcal{F})$ to $H^1(\mathcal{V},\mathcal{F})$.

Proposition 1. *If $\tau,\sigma : A \to I$ are two refinement maps (i.e. $V_\alpha \subset U_{\tau(\alpha)} \cap U_{\sigma(\alpha)}$ $\forall \alpha \in A$), then the induced maps*

$$\tau^*,\sigma^* : H^1(\mathcal{U},\mathcal{F}) \longrightarrow H^1(\mathcal{V},\mathcal{F})$$

are equal.

Proof. If $(f_{ij})_{i,j\in I} \in Z^1(\mathcal{U},\mathcal{F})$, we have, on $V_\alpha \cap V_\beta$,

$$f_{\tau(\alpha)\tau(\beta)} - f_{\sigma(\alpha)\sigma(\beta)} = \big(f_{\tau(\alpha)\sigma(\alpha)} + f_{\sigma(\alpha)\tau(\beta)}\big) - \big(f_{\sigma(\alpha)\tau(\beta)} + f_{\tau(\beta)\sigma(\beta)}\big) = g_\alpha - g_\beta\ ,$$

where $g_\alpha = f_{\tau(\alpha)\sigma(\alpha)} \mid V_\alpha$. Hence $\{f_{\tau(\alpha)\tau(\beta)} - f_{\sigma(\alpha)\sigma(\beta)}\} \in B^1(\mathcal{V},\mathcal{F})$.

Proposition 2. *If $\mathcal{V}$ is a refinement of $\mathcal{U}$, then, the induced map*

$$H^1(\mathcal{U},\mathcal{F}) \longrightarrow H^1(\mathcal{V},\mathcal{F})$$

is injective.

Proof. Let $\tau : A \to I$ be a refinement map, and let $\xi = \{(f_{ij})_{i,j\in I}\} \in Z^1(\mathcal{U},\mathcal{F})$, and suppose that $\tau^*(\xi) \in B^1(\mathcal{V},\mathcal{F})$. Thus, $\exists g_\alpha \in \mathcal{F}(V_\alpha)$ with $f_{\tau(\alpha)\tau(\beta)} \mid V_\alpha \cap V_\beta = g_\alpha|V_\alpha \cap V_\beta - g_\beta| V_\alpha \cap V_\beta$. Let $i \in I$ and $x \in U_i$. Choose $\alpha \in A$ so that $x \in V_\alpha$, and define $h_i(x) = g_\alpha(x) + f_{i\tau(\alpha)}(x)$. If $\beta \in A$ is such that $x \in V_\beta$, we have

$$g_\alpha(x) + f_{i\tau(\alpha)}(x) - g_\beta(x) - f_{i\tau(\beta)}(x) = g_\alpha(x) - g_\beta(x) - f_{\tau(\alpha)\tau(\beta)}(x) = 0$$

because

$$\begin{aligned} f_{i\tau(\alpha)}(x) - f_{i\tau(\beta)}(x) &= -\big(f_{\tau(\alpha)i}(x) + f_{i\tau(\beta)}(x)\big) \\ &= -f_{\tau(\alpha)\tau(\beta)}(x) \qquad (\text{since } \xi \in Z^1(\mathcal{U},\mathcal{F}))\ . \end{aligned}$$

Hence, the above formula defines $h_i \in \mathcal{F}(U_i)$.

If $x \in U_i \cap U_j$, and we choose α with $x \in V_\alpha$, we have

$$\begin{aligned} h_i(x) - h_j(x) &= g_\alpha(x) + f_{i\tau(\alpha)}(x) - g_\alpha(x) - f_{j\tau(\alpha)}(x) \\ &= f_{i\tau(\alpha)}(x) + f_{\tau(\alpha)j}(x) = f_{ij}(x)\ . \end{aligned}$$

Thus $\delta\{(h_i)\} = \xi$, and $\xi \in B^1(\mathcal{U}, \mathcal{F})$.

This proves Proposition 2.

We now define the cohomology group $H^1(X, \mathcal{F})$. Let $\mathcal{U}, \mathcal{V}$ be open coverings of X, $\mathcal{U} = \{U_i\}_{i \in I}$, $\mathcal{V} = \{V_\alpha\}_{\alpha \in A}$, $\mathcal{V}$ a refinement of $\mathcal{U}$. Then, there is a map $\tau(\mathcal{U}, \mathcal{V}) : H^1(\mathcal{U}, \mathcal{F}) \to H^1(\mathcal{V}, \mathcal{F})$ (defined using a refinement map $\tau : A \to I$, but independent of the choice of this map). If $\mathcal{W}$ is a refinement of $\mathcal{V}$, we have $\tau(\mathcal{U}, \mathcal{W}) = \tau(\mathcal{V}, \mathcal{W}) \circ \tau(\mathcal{U}, \mathcal{V})$. We define $H^1(X, \mathcal{F})$ as the direct limit of the system $\big(H^1(\mathcal{U}, \mathcal{F}), \tau(\mathcal{U}, \mathcal{V})\big)$ which is the following:

Let R be the equivalence relation on the disjoint union $\coprod_{\mathcal{U}} H^1(\mathcal{U}, \mathcal{F})$ defined by: $\xi \in H^1(\mathcal{U}, \mathcal{F})$ is equivalent to $\eta \in H^1(\mathcal{V}, \mathcal{F})$ if there is an open covering $\mathcal{W}$ which is a refinement of both $\mathcal{U}$ and of $\mathcal{V}$ and such that

$$\tau(\mathcal{U}, \mathcal{W})\xi = \tau(\mathcal{V}, \mathcal{W})\eta \,.$$

Then, $H^1(X, \mathcal{F}) = \coprod H^1(\mathcal{U}, \mathcal{F})/R$.
For any $\mathcal{U}$, there is a map $\tau(\mathcal{U}) : H^1(\mathcal{U}, \mathcal{F}) \to H^1(X, \mathcal{F})$ $\big(\tau(\mathcal{U})\xi$ = equivalence class of ξ under $R\big)$.

Proposition 2 is equivalent to the statement that $\tau(\mathcal{U})$ is injective.

We shall need the following special case of a theorem of Leray.

LERAY'S THEOREM. Let $\mathcal{F}$ be a sheaf of abelian groups on the topological space X. Let $\mathcal{U} = \{U_i\}_{i \in I}$ be an open covering of X. Suppose that $H^1(U_i, \mathcal{F}) = 0\ \forall i$. Then, the natural map

$$H^1(\mathcal{U}, \mathcal{F}) \longrightarrow H^1(X, \mathcal{F})$$

is an isomorphism.

Proof. It is sufficient (because of Proposition 2), to show that for any refinement $\mathcal{V} = \{V_\alpha\}_{\alpha \in A}$ of $\mathcal{U}$, the induced map $\tau^* : H^1(\mathcal{U}, \mathcal{F}) \to H^1(\mathcal{V}, \mathcal{F})$ is surjective; here $\tau : A \to I$ is a map with $V_\alpha \subset U_{\tau(\alpha)} \forall \alpha$.

Let $\{c_{\alpha\beta}\}_{\alpha,\beta \in A} \in Z^1(\mathcal{V}, \mathcal{F})$. Now $\{c_{\alpha,\beta}|U_i\} \in Z^1(U_i \cap \mathcal{V}, \mathcal{F})$, where $U_i \cap \mathcal{V}$ is the covering $\{U_i \cap V_\alpha\}_{\alpha \in A}$ of U_i. Since $H^1(U_i, \mathcal{F}) = 0$ by hypothesis, Proposition 1 implies that there exist $g_{i\alpha} \in \mathcal{F}(U_i \cap V_\alpha)$ such that

$$c_{\alpha\beta} = g_{i\alpha} - g_{i\beta} \quad \text{on} \quad U_i \cap V_\alpha \cap V_\beta \,.$$

Now, on $U_i \cap U_j \cap V_\alpha \cap V_\beta$ we have $g_{i\alpha} - g_{i\beta} = c_{\alpha\beta} = g_{j\alpha} - g_{j\beta}$, i.e. $g_{i\alpha} - g_{j\alpha} = g_{i\beta} - g_{j\beta}$; hence (by the 2nd sheaf axiom), there exist elements $\gamma_{ij} \in \mathcal{F}(U_i \cap U_j)$ so that $\gamma_{ij} = g_{i\alpha} - g_{j\alpha}$ on $U_i \cap U_j \cap V_\alpha$. Clearly, $\gamma_{ij} + \gamma_{ik} = \gamma_{ik}$ on $U_i \cap U_j \cap U_k$. We have, on $V_\alpha \cap V_\beta$ $(\subset U_{\tau(\beta)})$

$$\begin{aligned}\gamma_{\tau(\alpha)\tau(\beta)} + c_{\alpha\beta} &= +\big(g_{\tau(\alpha)\alpha} - g_{\tau(\beta)\alpha}\big) + \big(g_{\tau(\beta)\alpha} - g_{\tau(\beta)\beta}\big) \\ &= -g_{\tau(\beta)\beta} + g_{\tau(\alpha)\alpha} \,.\end{aligned}$$

Since $g_{\tau(\beta)\beta} \in \mathcal{F}(U_{\tau(\beta)} \cap V_\beta) = \mathcal{F}(V_\beta)$, this shows that $\{c_{\alpha,\beta}\}$ and $\{-\gamma_{\tau(\alpha)\tau(\beta)}\}$ induce the same element in $H^1(\mathcal{V}, \mathcal{F})$ but the latter is the image, under τ^*, of $\{-\gamma_{ij}\}$. This proves the theorem.

The relevance and usefulness of Leray's theorem in the theory of Riemann surfaces stems from the following:

MITTAG–LEFFLER'S THEOREM. Let Ω be an open set in $\mathbb{C}$. Then $H^1(\Omega, \mathcal{O}) = 0$, where $\mathcal{O}$ is the sheaf of germs of holomorphic functions on Ω.

To prove this, we first prove the following result.

Proposition 3. *Let Ω be open in $\mathbb{C}$ and let $f \in C^\infty(\Omega)$. Then, $\exists u \in C^\infty(\Omega)$ such that $\frac{\partial u}{\partial \bar{z}} = f$.*

Recall that, if $z = x + iy$, x, y real, then

$$\frac{\partial}{\partial \bar{z}} = \frac{1}{2}\left(\frac{\partial}{\partial x} + i\frac{\partial}{\partial y}\right), \quad \frac{\partial}{\partial z} = \frac{1}{2}\left(\frac{\partial}{\partial x} - i\frac{\partial}{\partial y}\right).$$

Proof of Proposition 3. Case 1. Suppose that f has compact support in Ω, and define

$$u(z) = \frac{1}{2\pi i}\int_{\mathbb{C}} \frac{f(z+w)}{w}\, dw \wedge d\bar{w}\,.$$

Then $u \in C^\infty(\mathbb{C})$ and $\frac{\partial u}{\partial \bar{z}} = f$. First, to see that $u \in C^\infty(\mathbb{C})$, remark that $\frac{1}{|w|}$ is integrable on any compact set in $\mathbb{C}$, (polar coordinates at 0), so that, for instance, the existence and continuity of $\frac{\partial u}{\partial x}$ follows from the fact that

$$\lim_{\substack{h\to 0\\ h\in\mathbb{R}}} \int_{\mathbb{C}} \frac{f(z+h+w) - f(z+w)}{h}\frac{1}{w}\, dw \wedge d\bar{w} = \int_{\mathbb{C}} \frac{\partial f}{\partial x}(z+w)\frac{1}{w}\, dw \wedge d\bar{w}$$

since, f having compact support, $\lim_{h\to 0} \frac{f(z+h)-f(z)}{h} = \frac{\partial f}{\partial x}(z)$, uniformly and boundedly on $\mathbb{C}$. We have only to iterate this argument. If $\varepsilon > 0$,

$$\frac{\partial u}{\partial \bar{z}} = \frac{1}{2\pi i}\lim_{\varepsilon\to 0} \int\limits_{|w|\geq\varepsilon} \frac{\partial f}{\partial \bar{z}}(z+w)\frac{1}{w}\, dw \wedge d\bar{w}\,;$$

now

$$\int\limits_{|w|\geq\varepsilon} \frac{\partial f}{\partial \bar{z}}(z+w)\frac{1}{w}\, dw \wedge d\bar{w} = \int\limits_{|w|\geq\varepsilon} \frac{\partial}{\partial \bar{w}}\left(\frac{f(z+w)}{w}\right) dw \wedge d\bar{w}$$

$$= -\int\limits_{|w|\geq\varepsilon} d\left(\frac{f(z+w)dw}{w}\right)$$

and by Stokes' theorem, this

$$= \int_{|w|=\varepsilon} \frac{f(z+w)}{w}\,dw = 2\pi i\, f(z) + \int_{|w|=\varepsilon} \frac{f(z+w)-f(z)}{w}\,dw\,.$$

Since $\frac{f(z+w)-f(z)}{w}$ is bounded as a function of w, this last integral $\to 0$ as $\varepsilon \to 0$, and the result follows.

If now $f \in C^\infty(\Omega)$, if we apply this special case to the function φf where $\varphi \in C_0^\infty(\Omega)$ and $= 1$ on a given compact set $K \subset \Omega$, we obtain the following:
If $f \in C^\infty(\Omega)$ and $K \subset \Omega$ is compact, there is $u \in C^\infty(\Omega)$ such that $\frac{\partial u}{\partial \bar z} = f$ on K.

To complete the proof of Proposition 3, we need the following form of Runge's theorem; we shall not prove this here. A proof is given e.g. in [7].

Let Ω be open in $\mathbb{C}$ and let $K \subset \Omega$ be compact. Let L be the union of K with those connected components of $\Omega - K$ which are relatively compact in Ω. Then L is compact and has the following property: any function holomorphic in a neighbourhood of L can be approximated, uniformly on L, by functions holomorphic on Ω.

To prove Proposition 3, let $\{K_n\}_{n\geq 1}$ be a sequence of compact sets in Ω with $K_n \subset \mathring{K}_{n+1}$ (interior of K_{n+1}), $\bigcup K_n = \Omega$ and such that $\Omega - K_n$ has no connected component relatively compact in Ω.

Let $f \in C^\infty(\Omega)$ and let $u_n \in C^\infty(\Omega)$ be such that $\frac{\partial u_n}{\partial \bar z} = f$ on a neighbourhood of K_n. Then $u_{n+1} - u_n$ is holomorphic on a neighbourhood of K_n, so that there is h_n, holomorphic on Ω so that $|u_{n+1} - u_n - h_n| < 2^{-n}$ on K_n ($n \geq 1$).

Define $u = u_n + \sum_{m\geq n}(u_{m+1} - u_m - h_m) - h_1 - \cdots - h_{n-1}$ on K_n; the series converges uniformly on K_n. We have

$$\begin{aligned} u &= u_n + (u_{n+1} - u_n - h_n) + \sum_{m\geq n+1} (u_{m+1} - u_m - h_m) - h_1 - \cdots - h_{n-1} \\ &= u_{n+1} + \sum_{m\geq n+1} (u_{m+1} - u_m - h_m) - h_1 - \cdots - h_n\,, \end{aligned}$$

so that this defines a function on Ω. Since $\sum_{m\geq n+1}(u_{m+1} - u_m - h_m)$ is holomorphic on $\mathring{K}_{n+1} \supset K_n$ and $\frac{\partial u_{n+1}}{\partial \bar z} = f$ on K_n, we have $\frac{\partial u}{\partial \bar z} = f$ on K_n $\forall n$, i.e. on Ω.

Proof of Mittag–Leffler's theorem. We shall prove that $H^1(\mathcal{U}, \mathcal{O}) = 0$ for any open covering $\mathcal{U}$ of Ω (Ω open in $\mathbb{C}$).

Given $\mathcal{U} = \{U_i\}_{i\in I}$, let $\{\alpha_i\}_{i\in I}$ be a partition of unity with respect to $\mathcal{U}$, i.e. $\alpha_i \in C^\infty(\Omega)$, support $(\alpha_i) \subset U_i$, the family $\{\text{support } (\alpha_i)\}_{i\in I}$ is locally finite, and $\sum \alpha_i \equiv 1$.

If c_{ij} is holomorphic on $U_i \cap U_j$ and $c_{ij} + c_{jk} = c_{ik}$ on $U_i \cap U_j \cap U_k$ for all i, j, k, taking $i = j = k$ we find that $c_{ii} = 0$ $\forall i$, now taking $k = i$, we find that $c_{ij} = -c_{ji}$.

Now, given i, j, we define a C^∞-function on U_i by setting it $= \alpha_j c_{ij}$ on $U_i \cap U_j$, $= 0$ on $U_i - (U_i \cap U_j)$. Since support $(\alpha_j) \subset U_j$, $\alpha_j = 0$ in a neighbourhood of $(\partial U_j) \cap U_i$, so that this function is C^∞ on U_i; we denote it simply by $\alpha_j c_{ij}$. Set $\varphi_i = \sum_{j \in I} \alpha_j c_{ij}$; this is C^∞ on U_i since {support (α_j)} is locally finite. Given $k, \ell \in I$, we have, on $U_k \cap U_\ell$,

$$\varphi_k - \varphi_\ell = \sum_j \alpha_j (c_{kj} - c_{\ell j}) = \sum_j \alpha_j (c_{kj} + c_{j\ell}) = \sum_j \alpha_j c_{k\ell} = c_{k\ell}\,.$$

Now, $\frac{\partial \varphi_k}{\partial \bar z} - \frac{\partial \varphi_\ell}{\partial \bar z} = \frac{\partial c_{k\ell}}{\partial \bar z} = 0$ on $U_k \cap U_\ell$. Thus, there is a C^∞-function ψ on Ω with $\psi | U_k = \frac{\partial \varphi_k}{\partial \bar z}$ $\forall k \in I$.

Let $u \in C^\infty(\Omega)$ and $\frac{\partial u}{\partial \bar z} = \psi$ on Ω, and let $h_k = \varphi_k - u$ on U_k. We have $\frac{\partial h_k}{\partial \bar z} = \frac{\partial \varphi_k}{\partial \bar z} - \frac{\partial u}{\partial \bar z} = 0$ on U_k, so that $h_k \in \mathcal{O}(U_k)$. On $U_k \cap U_\ell$, we have $h_k - h_\ell = \varphi_k - \varphi_\ell = c_{k\ell}$. This proves the theorem.

One more general construction and theorem will be required from sheaf theory. This is the *exact cohomology sequence*, and we shall develop just the part of this sequence which will be needed.

Let $0 \to \mathcal{E} \xrightarrow{\alpha} \mathcal{F} \xrightarrow{\beta} \mathcal{G} \to 0$ be a short exact sequence of sheaves of abelian groups on the topological space X. (Recall that the exactness at $\mathcal{G}$, i.e. the surjectivity of β amounts to the surjectivity of $\beta_x : \mathcal{F}_x \to \mathcal{G}_x$ $\forall x \in X$, where $\mathcal{F}_x$, $\mathcal{G}_x$ are the stalks of $\mathcal{F}$ and $\mathcal{G}$ at x.) We have

Lemma. *The induced sequence*

$$0 \longrightarrow \mathcal{E}(X) \xrightarrow{\alpha_X} \mathcal{F}(X) \xrightarrow{\beta_X} \mathcal{G}(X)$$

is exact.

Proof. Given the injectivity of $\alpha_x : \mathcal{E}_x \to \mathcal{F}_x$, the injectivity of α_X is simply the first of the sheaf axioms for $\mathcal{E}$.

The exactness at $\mathcal{F}(X)$ asserts that if $f \in \mathcal{F}(X)$ and there is a covering $\{U_i\}$ of X such that $f|U_i \in \operatorname{im}(\alpha_{U_i})$ $\forall i$, then $f \in \operatorname{im}(\alpha_X)$. Let $e_i \in \mathcal{E}(U_i)$ be such that $\alpha_{U_i}(e_i) = f|U_i$. We have $(e_i - e_j)|U_i \cap U_j = 0$, (since α is injective). By the second sheaf axiom, there is $e \in \mathcal{F}(X)$ with $e|U_i = e_i \forall i$. Clearly $\alpha_X(e)|_{U_i} = \alpha_{V_i}(e_i) = f|U_i$, so that $\alpha_X(e) = f$.

Recall also that $H^0(X, \mathcal{E}) = \mathcal{E}(X), \ldots$. We define $\alpha^0 : H^0(X, \mathcal{E}) \to H^0(X, \mathcal{F})$ by $\alpha^0 = \alpha_X$ (and similarly, $\beta^0 : H^0(X, \mathcal{F}) \to H^0(X, \mathcal{G})$ by $\beta^0 = \beta_X$). We now define morphisms $\delta : H^0(X, \mathcal{G}) \to H^1(X, \mathcal{E})$ and $\alpha^1 : H^1(X, \mathcal{E}) \to H^1(X, \mathcal{F})$, $\beta^1 : H^1(X, \mathcal{F}) \to H^1(X, \mathcal{G})$ as follows.

Definition of δ. Let $g \in H^0(X, \mathcal{G}) = \mathcal{G}(X)$. Since β is surjective, there is an open covering $\{U_i\}_{i \in I} = \mathcal{U}$ of X and elements $f_i \in \mathcal{F}(U_i)$ such that $\beta_{U_i}(f_i) = g_i|U_i$ $\forall i$. We have $f_i - f_j|U_i \cap U_j \in \ker(\beta_{U_i \cap U_j}) = \operatorname{im}(\alpha_{U_i \cap U_j})$ by the lemma above, i.e. $\exists e_{ij} \in \mathcal{E}(U_i \cap U_j)$ with $\alpha_{U_i \cap U_j}(e_{ij}) = f_i - f_j|U_i \cap U_j$. Clearly $e_{ij} + e_{jk} - e_{ik}$ maps to 0 under α, so is 0 on $U_i \cap U_j \cap U_k$. Thus $\{e_{ij}\} \in Z^1(\mathcal{U}, \mathcal{E})$, so defines an element $\delta(g) \in H^1(X, \mathcal{E})$. We have to check that this element is independent of the choices: viz of $\mathcal{U}$ and the $\{f_i\}$. If $\mathcal{V} = \{V_\alpha\}_{\alpha \in A}$ is a refinement of $\mathcal{U}$ (given by $\tau : A \to I$) and we choose the elements $f'_\alpha = f_{\tau(\alpha)}|V_\alpha$, we clearly get the same element of $H^1(X, \mathcal{E})$ (in fact simply the restriction of the cocycle $\{e_{ij}\}$ defined above to $\mathcal{V}$).

Hence we consider the same covering $\mathcal{U}$ and possibly different liftings $f'_i \in \mathcal{F}(U_i)$ with $\beta_{U_i}(f'_i) = g|U_i$. As above, we see that there are elements $e_i \in \mathcal{E}(U_i)$ such that $\alpha_{U_i}(e_i) = f'_i - f_i$. If $e_{ij} \in \mathcal{E}(U_i \cap U_j)$ map to $f'_i - f'_j$ on $U_i \cap U_j$, we have $e_i - e_j = e'_{ij} - e_{ij}$ (since the image under α is 0), and $\{e_{ij}\}, \{e'_{ij}\}$ define the same element of $H^1(X, \mathcal{E})$.

Definition of α^1. The map $\alpha : \mathcal{E} \to \mathcal{F}$ induces, for any covering $\mathcal{U}$ of X, a map

$$\alpha_{\mathcal{U}} : C^1(\mathcal{U}, \mathcal{E}) \longrightarrow C^1(\mathcal{U}, \mathcal{F})$$

which maps cocycles into cocycles and coboundaries to coboundaries, hence induces a morphism $\alpha^1_{\mathcal{U}} : H^1(\mathcal{U}, \mathcal{E}) \to H^1(\mathcal{U}, \mathcal{F})$. This in turn induces α^1.

Definition of β^1. This is the map $H^1(X, \mathcal{F}) \to H^1(X, \mathcal{G})$ induced by $\beta : \mathcal{F} \to \mathcal{G}$ (as with α above).

Theorem *(The Exact Cohomology Sequence).*

Let $0 \to \mathcal{E} \xrightarrow{\alpha} \mathcal{F} \xrightarrow{\beta} \mathcal{G} \to 0$ *be a short exact sequence of sheaves on the topological space* X. *Then, the sequence*

$$0 \longrightarrow H^0(X, \mathcal{E}) \xrightarrow{\alpha^0} H^0(X, \mathcal{F}) \xrightarrow{\beta^0} H^0(X, \mathcal{G}) \xrightarrow{\delta}$$

$$\xrightarrow{\delta} H^1(X, \mathcal{E}) \xrightarrow{\alpha^1} H^1(X, \mathcal{F}) \xrightarrow{\beta^1} H^1(X, \mathcal{G})$$

is exact.

Remark. One can define Čech cohomology groups $H^q(X, \mathcal{F})$ for all $q \geq 0$ and extend the exact sequence above when X is paracompact [see e.g. Serre–Faisceaux algébriques cohérents, Annals of Math. 61(1955)].

Proof. Exactness at $H^0(X, \mathcal{G})$. First, if $g = \beta^0(f)$, $f \in H^0(X, \mathcal{G})$, then, in the definition of $\delta(g)$, we can take any covering $\{U_i\}$ and $f_i = f|U_i$. Since $f_i - f_j|U_i \cap U_j = 0$, we have $\delta(g) = 0$.

Conversely, suppose that $\delta(g) = 0$; it is represented, with respect to a suitable covering $\mathcal{U} = \{U_i\}_{i \in I}$ by $\{e_{ij}\} \in Z^1(\mathcal{U}, \mathcal{E})$ with $\alpha_{U_i \cap U_j}(e_{ij}) = f_i - f_j|U_i \cap U_j$ and $\beta_{U_i}(f_i) = g|U_i$. Since $\delta(g) = 0$ (and $H^1(\mathcal{U}, \mathcal{E}) \to H^1(X, \mathcal{E})$ is injective), there are elements $e_i \in \mathcal{E}(U_i)$

with $e_i - e_j|U_i \cap U_j = e_{ij}$. Let $f_i' = f_i - \alpha_{U_i}(e_i)$; then $f_i' - f_j' = f_i - f_j - \alpha_{U_i \cap U_j}(e_{ij}) = 0$, so that there is $f \in \mathcal{F}(X)$ with $f|U_i = f_i$. We have $\beta_{U_i}(f) = g|U_i - (\beta \circ \alpha)_{U_i}(e_i) = g|U_i$; thus $g = \beta_X(f)$.

Exactness at $H^1(X, \mathcal{E})$. Let $\mathcal{U} = \{U_i\}$ be a covering and $\{e_{ij}\} \in Z^1(\mathcal{U}, \mathcal{E})$. Then $\alpha^1(\xi) = 0$ [ξ being the class in $H^1(X, \mathcal{E})$ of $\{e_{ij}\}$] $\Longleftrightarrow \exists \{f_i\} \in C^0(\mathcal{U}, \mathcal{F})$ with $f_i - f_j|U_i \cap U_j = \alpha_{U_i \cap U_j}(e_{ij})$; this condition is, by definition, satisfied if $\xi = \delta(g)$. Conversely, if this holds, then $f_i - f_j|U_i \cap U_j \in \ker(\beta)$, hence there is $g \in H^0(X, \mathcal{G})$ with $g|U_i = \beta_{U_i}(f_i)$, and, again by definition, $\xi = \delta(g)$.

Exactness at $H^1(X, \mathcal{F})$. We could define a map $(\beta \circ \alpha)^1 : H^1(X, \mathcal{E}) \to H^1(X, \mathcal{G})$ induced by $\beta \circ \alpha = 0$ as we defined α^1; clearly $(\beta \circ \alpha)^1 = 0$. But we see at once that $(\beta \circ \alpha)^1 = \beta^1 \circ \alpha^1$, so that $\mathrm{im}(\alpha^1) \subset \ker(\beta^1)$.

Conversely, let $\{f_{ij}\} \in Z^1(\mathcal{U}, \mathcal{F})$ [$\mathcal{F}$ a suitable covering of X] and suppose that $\beta(f_{ij}) = g_i - g_j|U_i \cap U_j$ where $g_i \in \mathcal{G}(U_i)$. By shrinking the covering $\mathcal{U}$, we may suppose (since β is surjective) that $\exists f_i \in \mathcal{F}(U_i)$ with $\beta(f_i) = g_i$. But then, if $f_{ij} - (f_i - f_j)|U_i \cap U_j = f_{ij}'$, $\{f_{ij}'\}$ represents the same element of $H^1(X, \mathcal{F})$ as $\{f_{ij}\}$, and $f_{ij}' \in \ker \beta_{U_i \cap U_j} = \mathrm{im}\, \alpha_{U_i \cap U_j}$. If $\alpha_{U_i \cap U_j}(e_{ij}) = f_{ij}'$, we see at once that $\{e_{ij}\} \in Z^1(\mathcal{U}, \mathcal{E})$ and its image under α^1 is the class of $\{f_{ij}'\}$ = class of $\{f_{ij}\}$.

6. Vector Bundles, Line Bundles and Divisors

Let X be a topological space. Suppose given another topological space E and a continuous map $\pi : E \to X$, and that each fibre $\pi^{-1}(a) = E_a$, $a \in X$, is provided with the structure of a $\mathbb{C}$-vector space of dimension n.

We call $\pi : E \to X$ a *(continuous) vector bundle* if π is locally trivial in the following sense:

$\forall a \in X$, $\exists$ an open neighbourhood U of a and a homeomorphism $h_U : \pi^{-1}(U) \to U \times \mathbb{C}^n$ with the following properties:

(i) The diagram

$$\begin{array}{ccc} \pi^{-1}(U) & \xrightarrow{h_U} & U \times \mathbb{C}^n \\ \pi \searrow & & \swarrow pr_U \\ & U & \end{array}$$

(where $pr_U : U \times \mathbb{C}^n$ is projection on the first factor) commutes.

(ii) $\forall a \in U$, the map $\varphi_a : E_a \to \mathbb{C}^n$ defined by $h_U(x) = (a, \varphi_a(x))$, $x \in E_a$, is an isomorphism of $\mathbb{C}$-vector spaces.

If X, E, π are C^∞ (resp. complex analytic) and if h_U can be chosen also C^∞ (resp. biholomorphic), we shall speak of C^∞ (resp. holomorphic) vector bundles. The integer n is called the *rank* of the vector bundle. If the rank $n = 1$, we call $\pi : E \to X$ a *line bundle*. The map h_U is called a trivialisation of E on U (or a local trivialisation of E at a). It is also sometimes called a linear chart for E on U.

If $\pi : E \to X$ is a holomorphic vector bundle on the complex manifold X, we can find an open covering $\mathcal{U} = \{U_i\}_{i \in I}$ and holomorphic trivialisations

$$h_i : \pi^{-1}(U_i) \longrightarrow U_i \times \mathbb{C}^n .$$

On $U_i \cap U_j$, the map $h_i \circ h_j^{-1} : (U_i \cap U_j) \times \mathbb{C}^n \longrightarrow (U_i \cap U_j) \times \mathbb{C}^n$ has the form $(x, v) \mapsto (x, \eta(x, v))$, where, for fixed $x \in U_i \cap U_j$, $v \mapsto \eta(x, v)$ is a $\mathbb{C}$-linear isomorphism of $\mathbb{C}^n$. Thus, there is a holomorphic map $g_{ij} : U_i \cap U_j \to GL(n, \mathbb{C})$ such that

$$h_i \circ h_j^{-1}(x, v) = (x, g_{ij}(x)v) .$$

On $U_i \cap U_j \cap U_k$, we have the cocycle condition

$$g_{ij}(x) \cdot g_{jk}(x) = g_{ik}(x)$$

(multiplication in $GL(n, \mathbb{C})$). The $\{g_{ij}\}$ are called transition functions of the vector bundle (corresponding to the local trivialisations h_i).

If we replace $\{h_i\}$ by other trivialisations $\{h_i'\}$, then $h_i' \circ h_i^{-1} : U_i \times \mathbb{C}^n$ is of the form $(x, v) \mapsto ((x, \varphi_i(x)v)$, where $\varphi_i : U_i \to GL(n, \mathbb{C})$ is holomorphic; the corresponding transition functions are $\varphi_i\, g_{ij}\, \varphi_j^{-1}$.

Conversely, given a family of holomorphic maps $g_{ij} : U_i \cap U_j \to GL(n, \mathbb{C})$ satisfying the cocycle condition $g_{ij}\, g_{jk} = g_{ik}$ on $U_i \cap U_j \cap U_k$, we can construct a vector bundle as follows.

Let $\tilde{E} = \coprod_{i \in I} U_i \times \mathbb{C}^n$ (disjoint union), $\tilde{\pi} : \tilde{E} \to X$ the map defined by $(x, v) \mapsto x$. Call $(x, v) \in U_i \times \mathbb{C}^n$ equivalent to $(y, w) \in U_j \times \mathbb{C}^n$ if $x = y$ and $v = g_{ij}(x)w$. This is an equivalence relation because of the cocycle condition, and two points in the same $U_i \times \mathbb{C}^n$ are equivalent only if they are equal. Moreover, $\tilde{\pi}$ induces a map $\pi : E \to X$ where E is the quotient of $\tilde{E}$ by the equivalence relation. The projection from $\tilde{E}$ to E induces a bijection h_i^{-1} from $U_i \times \mathbb{C}^n$ to $\pi^{-1}(U_i)$, and one checks easily that $\pi : E \to X$ is a holomorphic vector bundle.

If $\pi : E \to X$ is a vector bundle, a (continuous, C^∞, holomorphic...) section of E is a (continuous, C^∞, holomorphic...) map $s : X \to E$ such that $\pi \circ s =$ identity on X. If we consider local trivialisations $h_i : \pi^{-1}(U_i) \to U_i \times \mathbb{C}^n$ as above, then $h_i \circ s(x) = (x, f_i(x))$ for $x \in U_i$, where f_i is a map $U_i \to \mathbb{C}^n$. Since $h_i \circ s(x) = h_i \circ h_j^{-1}(x, f_j(x)) = (x, g_{ij}(x) f_j(x))$ if $x \in U_i \cap U_j$, we have

$$f_i(x) = g_{ij}(x) f_j(x)\ , x \in U_i \cap U_j\ .$$

The converse also holds, as is checked at once. Thus a (continuous, C^∞, holomorphic...) section of $\pi : E \to X$ can be identified with a family $\{f_i\}_{i \in I}$ of (continuous, C^∞, holomorphic...) maps $f_i : U_i \to \mathbb{C}^n$ such that $f_i = g_{ij} f_j$ on $U_i \cap U_j$. Sections of E over open subsets of X are defined in the obvious way.

Let now X be a Riemann surface and $\pi : E \to X$ a holomorphic vector bundle on X. A *meromorphic* section of E is defined as follows: let $S \subset X$ be a discrete set and let $s : X - S \to E$ be a holomorphic section. Then s is said to be a meromorphic section of E if, $\forall a \in S$, there is a neighbourhood U of a and a coordinate z on U with $z(a) = 0$ such that $U \cap S = \{a\}$ and for some integer $N \geq 0$, $z^N s$ is the restriction to $U - \{a\}$ of a holomorphic section of E over U.

If $\mathcal{U} = \{U_i\}_{i \in I}$ is an open covering of X and $h_i : \pi^{-1}(U_i) \to U_i \times \mathbb{C}^n$ are trivialisations, we can write $h_i \circ s(x) = (x, f_i(x))$ for $x \in U_i - S$; f_i is holomorphic on $U_i - S$, and the section s is meromorphic if and only if f_i is meromorphic on U_i for all i (i.e. if f_i has at most a pole at points of $U_i \cap S$).

If $\pi : E \to X$, $\pi' : E' \to X$ are vector bundles (continuous, C^∞, holomorphic...) a morphism $u : E \to E'$ is a map such that $\pi' \circ u = \pi$ and $u : \pi^{-1}(a) \to \pi'^{-1}(a)$ is $\mathbb{C}$-linear. It is called continuous, C^∞, holomorphic,... if u has this property. Two bundles

E, E' are isomorphic if there are morphisms $u : E \to E'$ and $u' : E' \to E$ such that $u \circ u' =$ identity on E' and $u' \circ u =$ identity on E. A bundle $\pi : E \to X$ is called trivial if it is isomorphic to the "trivial bundle" $pr_X : X \times \mathbb{C}^n \to X$ (where $pr_X(x, v) = x$). Isomorphisms, and trivialisations, can be continuous,C^∞, holomorphic,... ; this simply means that, e.g. the morphisms $u : E \to E'$, $u' : E' \to E$ have this property.

We proceed now to define divisors on a Riemann surface. Let X be a Riemann surface. A divisor D on X is a map $D : X \to \mathbb{Z}$ such that the support of D is locally finite, i.e. $\forall K \subset X$ compact, the set $\{P \in X | D(P) \neq 0\}$ is finite. We usually write

$$D = \sum_{P \in X} D(P)P \, ;$$

if X is compact, the sum is finite. We define the sum and difference $D_1 \pm D_2$ of divisors D_1, D_2 by $(D_1 \pm D_2)(P) = D_1(P) \pm D_2(P)$. We say that a divisor D is *effective* if $D(P) \geq 0 \; \forall P \in X$. Given divisors D_1, D_2, we say $D_1 \geq D_2$ if $D_1 - D_2$ is effective. The set $\{P \in X | D(P) \neq 0\}$ is called the support of D and written $\operatorname{supp}(D)$.

Recall the following notation: if f is meromorphic in a neighbourhood U of a point a on a Riemann surface X and z is a local coordinate on U with $z(a) = 0$, we set

$$\operatorname{ord}_a(f) = \begin{cases} \infty & \text{if } f \equiv 0 \text{ near } a \\ k & \text{if } f(z) = \sum_{n=k}^{\infty} a_n z^n \quad (k \in \mathbb{Z}), \; a_k \neq 0 \, . \end{cases}$$

Let now s be a meromorphic section of a holomorphic vector bundle E on a Riemann surface X, $s \not\equiv 0$.

For $a \in X$, choose a coordinate neighbourhood (U, z) with $z(a) = 0$ and a trivialisation $h : E|U \to U \times \mathbb{C}^n$. Then $h \circ s(x) = \big(x, f(x)\big)$, $x \in U$, where f is an n-tuple of meromorphic functions on U. There is an integer k such that $f = z^k g$, where g is an n-tuple of functions holomorphic near a, and $g(a) \neq 0$. We set $k = \operatorname{ord}_a(s)$.

One checks easily that this is independent of the local coordinate and of the local trivialisation.

If s is a meromorphic section of a holomorphic vector bundle E, the divisor $a \mapsto \operatorname{ord}_a(s)$, i.e. the divisor

$$\sum_{a \in X} \operatorname{ord}_a(s) a \, ,$$

is called the divisor of s and denoted by (s) or $\operatorname{div}(s)$. In fact, any divisor on a Riemann surface can be obtained as the divisor of a meromorphic section, even of a line bundle (i.e. vector bundle of rank 1), and we turn now to this construction.

Let X be a Riemann surface and D a divisor on X; let $D = \sum_{P \in X} n_P P$ (where $n_P = D(P) \in \mathbb{Z}$ and the sum is locally finite). Let $S = \{P \in X | n_P \neq 0\}$ and let $\{U_P, z_P\}$ be a local coordinate system at $P \in S$ with $z_P(P) = 0$; we assume that the open sets U_P are so chosen that $U_P \cap U_Q = \emptyset$ if $P \neq Q$, $P, Q \in S$. Let $U_* = X - S$, let $f_* \equiv 1$ and let $f_P = z_P^{n_P}$ on U_P, $P \in S$. If $I = \{*\} \coprod S$, let $g_{ij} = f_i/f_j$, $i, j \in I$, g_{ij} being defined on $U_i \cap U_j$ [$g_{ij} = 1$ by convention if $U_i \cap U_j = \emptyset$]. Now, if $U_i \cap U_j \neq \emptyset$, g_{ij} is holomorphic and nowhere 0 on this intersection. Moreover, $g_{ij}\, g_{jk} = g_{ik}$ on $U_i \cap U_j \cap U_k$ (if this latter intersection is $\neq \emptyset$). Thus, the $\{g_{ij}\}$ form a system of transition functions for a line bundle $L(D)$. Moreover, the functions $\{f_i\}_{i \in I}$ define a meromorphic section of $L(D)$ since, by definition, $f_i = g_{ij} f_j$, $i, j \in I$. We shall denote this section by s_D. Since s_D is defined by f_i on U_i, we have $\mathrm{ord}_a(s_D) = \mathrm{ord}_a(f_*) = 0$ if $a \in X - S$, and $\mathrm{ord}_a(s_D) = \mathrm{ord}_P(z_P^{n_P}) = n_P$ if $a = P \in S$. Thus $(s_D) = D$.

If we use a different local coordinate (U_P, ζ_P) at P (with the same U_P), then $h_P = (\zeta_P/z_P)^{n_P}$ is holomorphic and non-zero on U_P, and if we set $h_* = 1$, and denote by $\{g_{ij}\}$ the transition functions obtained from the functions $\zeta_P^{n_P}$, then $g'_{ij} = h_i g_{ij} h_j^{-1}$. If L' is the line bundle defined by the $\{g'_{ij}\}$, there is therefore an isomorphism of $L(D)$ onto L' taking s_D to the corresponding section of L' defined by the $\{f'_i\}$ ($f'_* = 1$, $f'_P = \zeta_P^{n_P}$).

Some general remarks. 1) If s is a meromorphic section $\not\equiv 0$ of a holomorphic vector bundle on X, then s is holomorphic if and only if $\mathrm{div}(s)$ is effective [s may have zeros, but no poles].

2) Given a divisor D, and $U \subset X$ open, let $\mathcal{O}_D(U) = \{f \text{ meromorphic on } U \mid \mathrm{div}(f) \geq -D \text{ on } U, \text{ i.e. } \mathrm{ord}_a(f) \geq -D(a)\ \forall a \in U\}$. The assignment $U \mapsto \mathcal{O}_D(U)$ is clearly a sheaf, denoted $\mathcal{O}_D$. If $\Gamma(U, L(D))$ denotes the space of holomorphic sections of $L(D)$ over U, and $s_D \in \Gamma(X, L(D))$ is the standard section with $\mathrm{div}(s_D) = D$, then the map $\mathcal{O}_D(U) \to \Gamma(U, L(D))$, $f \mapsto f s_D$, is an *isomorphism*, in fact $\mathrm{div}(f) \geq -D = -\mathrm{div}(s_D) \Longleftrightarrow \mathrm{div}(f s_D) \geq 0$.

Thus, *the sheaf of germs of holomorphic sections of $L(D)$ can be canonically identified with $\mathcal{O}_D$*.

We also remark that if $\pi : L \to X$ is a line bundle, and if s_0, s_1 are two meromorphic sections of L (with $s_0 \not\equiv 0$), then there is a meromorphic function f on X with $s_1 = f s_0$.

We now define *linear equivalence* of divisors. Two divisors D_1, D_2 are said to be linearly equivalent if there exists a meromorphic function f on X, $f \not\equiv 0$, such that

$$(f) = D_1 - D_2\ , \quad \text{we then write } D_1 \sim D_2\ .$$

Lemma. *Two divisors D_1, D_2 are linearly equivalent if and only if the line bundles $L(D_1)$ and $L(D_2)$ are (holomorphically) isomorphic.*

Proof. Suppose D_1 and D_2 are linearly equivalent, and let f be meromorphic on X with $(f) = D_1 - D_2$. Let s_{D_1}, s_{D_2} be the standard sections of $L(D_1)$, $L(D_2)$ respectively.

There is a unique isomorphism $u : L(D_1) \to L(D_2)$ taking s_{D_1} to fs_{D_2}; u is defined for x outside the zeros and poles of f and the supports of D_1 and D_2 by

$$\lambda s_{D_1}(x) \longmapsto \lambda f(x) s_{D_2}(x)\ ,\ \lambda \in \mathbb{C}\ ;$$

it extends holomorphically to X because, on any open set $U \subset X$, the section $x \mapsto \lambda(x)s_{D_1}(x)$ is holomorphic if and only if $x \mapsto \lambda(x)f(x)s_{D_2}(x)$ is holomorphic [$\mathrm{div}(s_{D_1}) \cap U = \mathrm{div}(fs_{D_2}) \cap U$].

Conversely, if $u : L(D_1) \to L(D_2)$ is an isomorphism, then $u \circ s_{D_1}$ is a meromorphic section of $L(D_2)$. If f is defined by $u \circ s_{D_1} = fs_{D_2}$, we have $(f) = D_1 - D_2$ (since $\mathrm{ord}_a(s_{D_1}) = \mathrm{ord}_a(u \circ s_{D_1})\ \forall a \in X$).

Remark. In terms of the sheaves $\mathcal{O}_{D_1}$ and $\mathcal{O}_{D_2}$, the isomorphism above is simply the map: $\varphi \in \mathcal{O}_{D_1}(U) \mapsto f\varphi \in \mathcal{O}_{D_2}(U)$.

Some further remarks. If $\pi : E \to X$ is a holomorphic vector bundle, and $E^* = \coprod_{a\in X}(E_a)^*$ (E_a^* is the dual of the vector space $E_a = \pi^{-1}(a)$), we can make E^* into a vector bundle in a natural way as follows. Let $U \subset X$ be open and $h_U : \pi^{-1}(U) \to U \times \mathbb{C}^n$ a trivialisation. We define $\check{h}_U : \coprod_{a\in U} E_a^* \to U \times \mathbb{C}^n$ by $\check{h}_U(v) = \left(a, (h_{U,a}^*)^{-1}(v)\right)$, $v \in E_a^*$, where $h_{U,a} : E_a \to \mathbb{C}^n$ is the isomorphism induced by $h_U \mid E_a \to \{a\} \times \mathbb{C}^n$ and $h_{U,a}^* : \mathbb{C}^n \to E_a^*$ is the dual map. If $g_{ij} : U_i \cap U_j \to GL(n, \mathbb{C})$ are the transition functions of E, those of E^* are ${}^t g_{ij}^{-1}$, (${}^tM$ denotes the transpose of a matrix M).

If E_1, E_2 are vector bundles on X, we define a vector bundle $\pi : E_1 \otimes E_2 \to X$ with $\pi^{-1}(a) = E_{1,a} \otimes E_{2,a}$; if $\{g_{ij}^{(\nu)}\} : U_i \cap U_j \to GL(n_\nu, \mathbb{C})$ are transition functions of E_ν^* ($\nu = 1, 2$), $E_1 \otimes E_2$ has transition functions $g_{ij}^{(1)} \otimes g_{ij}^{(2)}$ (Kronecker product). In particular, if L_1, L_2 are line bundles with transition functions $g_{ij}^{(1)}$, $g_{ij}^{(2)}$ respectively, $L_1 \otimes L_2$ has transition functions $g_{ij}^{(1)} \cdot g_{ij}^{(2)}$ (multiplication in $\mathbb{C}^* = \mathbb{C} - \{0\}$).

If D_1, D_2 are divisors, we see directly from the construction that $L(D_1) \otimes L(D_2)$ is isomorphic to $L(D_1 + D_2)$. Moreover, $L(-D)$ is isomorphic to $L(D)^*$ for any divisor D. If L is a line bundle and if it has a section s with $s(x) \neq 0\ \forall x \in X$, then L is trivial; in fact, the map $X \times \mathbb{C} \to L$ given by $(x, \lambda) \mapsto \lambda s(x)$ is an isomorphism. It follows that if L is a line bundle on X, then $L \otimes L^*$ is trivial: if $v \in L_x$, $v \neq 0$, there is a unique linear form ℓ on L_x such that $\ell(v) = 1$ [and the form corresponding to cv, $c \in \mathbb{C}$, is $\frac{1}{c}\ell$]. Thus $x \mapsto v \otimes \ell$ is a nowhere zero section of $L \otimes L^*$.

If L is a holomorphic line bundle on X and s is a meromorphic section of L, $s \not\equiv 0$, then $L \simeq L(D)$, where $D = \mathrm{div}(s)$. In fact, if s_{-D} is the standard section of $L(-D)$, then $s \otimes s_{-D}$ is a nowhere vanishing holomorphic section of $L \otimes L(-D)$. [Alternately, the isomorphism $\lambda s(x) \mapsto \lambda s_D(x)$ defined outside the support of D extends to an isomorphism of L onto $L(D)$.]

7. Finiteness Theorems

Theorem 1. *Let X be a compact Riemann surface and $\pi : E \to X$ a holomorphic vector bundle on X. Then the space $H^0(X, E)$ of global holomorphic sections of E on X : $H^0(X, E) = \{s : X \to E \mid \pi \circ s =$ identity on X, s holomorphic$\}$ is a finite dimensional vector space.*

Proof. We choose finitely many coordinate neighbourhoods $\{U_i, z_i\}_{i=1,\dots N}$ with the following properties:

(a) $z_i : U_i \to \Delta = \{z \in \mathbb{C} \mid |z| < 1\}$ is an analytic isomorphism.

(b) If $V_i = z_i^{-1}\{z \in \mathbb{C} \mid |z| < \frac{1}{2}\}$, then $\bigcup V_i = X$.

(c) There exist neighbourhoods W_i of $\bar{U}_i$ and trivialisations $h_i : \pi^{-1}(W_i) \to W_i \times \mathbb{C}^n$, with corresponding transition functions $g_{ij} : W_i \cap W_j \to GL(n, \mathbb{C})$.

Let $s \in H^0(X, E)$; then s is represented by a family $\{s_i\}_{i=1,\dots,N}$ of holomorphic maps $s_i : W_i \to \mathbb{C}^n$ such that

$$s_i(x) = g_{ij}(x)s_j(x) \quad \text{for} \quad x \in W_i \cap W_j \ .$$

We set

$$\|s\|^U = \max_i \sup_{x \in U_i} |s_i(x)|$$

and

$$\|s\|^V = \max_i \sup_{x \in V_i} |s_i(x)| \ .$$

First remark the following: There exists a constant $C > 0$ such that, $\forall s \in H^0(X, E)$, we have

$$\|s\|^U \leq C \|s\|^V \ .$$

In fact, let $x_0 \in \bar{U}_i$ be such that $|s_i(x_0)| = \|s\|^U$. Choose j such that $x_0 \in V_j$.

We have

$$|s_i(x_0)| = |g_{ij}(x_0)s_j(x_0)| \leq C\,|s_j(x_0)| \leq C \|s\|^V \ ,$$

where $C = \max_{i,j} \sup_{x \in U_i \cap U_j} \|g_{ij}(x)\|$, $\|g\|$ denoting the operator norm of $g \in GL(n, \mathbb{C})$ (considerd as a linear map of $\mathbb{C}^n$ into itself).

Let $a_i \in U_i$ be the point with $z_i(a_i) = 0$. We now prove the following (Schwarz's Lemma): Let $s \in H^0(X, E)$ and suppose that $\text{ord}_{a_i}(s) \geq k$ ($k \geq 0$ a given integer), $i =$

$1, \ldots, N$. Then $\|s\|^V \le 2^{-k} \|s\|^U$; in fact, $\frac{s_i}{z_i^k}$ is holomorphic in U_i, so that $\sup_{V_i} \left|\frac{s_i}{z_i^k}\right| \le \sup_{\partial U_i} \left|\frac{s_i}{z_i^k}\right| = \sup_{\partial U_i} |s_i| \le \|s\|^U$. Hence, if $x \in V_i$, $|s_i(x)| \le \sup_{z_i \in V_i} \left(|z_i^k| \left|\frac{s_i}{z_i^k}\right|\right) \le 2^{-k} \|s\|^U$. Hence, if $s \in H^0(X, E)$ and $\mathrm{ord}_{a_i}(s) \ge k$, we have

$$\|s\|^U \le C \|s\|^V \le 2^{-k} C \|s\|^U .$$

Thus, if $2^k > C$, it follows that $s \equiv 0$.

If $\mathcal{O}_{a_i}$ is the ring of germs of holomorphic functions at a_i and m_i^k, the ideal in $\mathcal{O}_{a_i}$ generated by z_i^k, then $\mathcal{O}_{a_i}/m_i^k$ is a $\mathbb{C}$-vector space of dimension k. Moreover, if $2^k > C$, the map

$$H^0(X, E) \longrightarrow \bigoplus_{i=1}^{n} \mathbb{C}^n \otimes (\mathcal{O}_{a_i}/m_i^k)$$

$$s \longmapsto \bigoplus_i (s_i \bmod z_i^k)$$

is injective, because the kernel consists exactly of sections s with $\mathrm{ord}_{a_i}(s) \ge k \ \forall i$. This proves the theorem.

The next finiteness theorem we shall need is somewhat more difficult to prove.

Let X be a compact Riemann surface and $\pi : E \to X$ a holomorphic vector bundle on X. The sheaf of germs of sections $\mathbb{E}$ of E is the sheaf $U \mapsto \mathbb{E}(U) = \{$space of holomorphic sections of E on $U\}$. We shall denote by $H^1(X, E)$ the first cohomology space $H^1(X, \mathbb{E})$.

Theorem 2. *If $\pi : E \to X$ is a holomorphic vector bundle on a compact Riemann surface X, the first cohomology $H^1(X, E)$ is a finite dimensional $\mathbb{C}$-vector space.*

Proof. Let $U \subset X$ be open, and suppose that there is a (holomorphic) trivialisation $h_U : \pi^{-1}(U) \to U \times \mathbb{C}^n$. Then, if V is open and $V \subset\subset U$ (relatively compact in U), we shall denote by $E_b(V)$ the space of *bounded* holomorphic sections of E on V, viz, the space of sections $s : V \to E$ such that if $h_U \circ s(x) = (x, f(x))$, $x \in V$, $f(x) \in \mathbb{C}^n$, then $\sup_{x \in V} |f(x)| < \infty$; we set $\|s\|_V = \sup_{x \in V} |f(x)|$. With this norm, $E_b(V)$ is a Banach space; a different trivialisation $h'_U : \pi^{-1}(U) \to U \times \mathbb{C}^n$ gives rise to an equivalent norm on $E_b(V)$.

If U, h_U are as above, and if, in addition, U is analytically isomorphic to an open set in $\mathbb{C}$, then $H^1(U, E) = 0$. This follows from the Mittag–Leffler theorem in §5 and the fact that if U is isomorphic to $\Omega \subset \mathbb{C}$ and $h_U : \pi^{-1}(U) \to U \times \mathbb{C}^n$ is an isomorphism, then $H^1(U, E) \simeq \bigoplus_{n \text{ copies}} H^1(\Omega, \mathcal{O})$.

Let $\Delta(r)$ be the disc $\{z \in \mathbb{C} \mid |z| < r\}$, $r > 0$. We choose a finite family $\{W_i, z_i\}_{i=1,\ldots,N}$ of coordinate neighbourhoods on X and holomorphic trivialisations $h_i : \pi^{-1}(W_i) \to W_i \times \mathbb{C}^n$ of E on W_i with the following properties:

1) z_i is an isomorphism of W_i onto $\Delta(2)$.
2) Setting $U_i(r) = z_i^{-1}(\Delta(r))$, we have $\bigcup_i U_i(\frac{1}{2}) = X$.

For $\frac{1}{2} \le r \le 2$, we denote by $\mathcal{U}(r)$ the covering $\{U_i(r)\}_{i=1,\dots,N}$ of X. Also, if $x \in W_i$ and $v \in \pi^{-1}(x) = E_x$, we write $|h_i(v)|$ for $|w|$ if $h_i(v) = (x, w)$, $w \in \mathbb{C}^n$. Set

$$Z_b^1(r) = \left\{\xi \in Z^1(\mathcal{U}(r), E) \mid \quad \text{if} \quad \xi = (f_{ij}), \quad \text{then} \quad f_{ij} \in E_b(U_i(r) \cap U_j(r)) \forall i, j\right\},$$

and

$$C_b^0(r) = \left\{\gamma \in C^0(\mathcal{U}(r), E) \mid \quad \text{if} \quad \gamma = (c_i). \quad \text{then} \quad c_i \in E_b(U_i(r)) \forall i\right\}.$$

We introduce norms $\| \ \|_r$ on these spaces by

$$\|\xi\|_r = \max_{i,j} \sup_{x \in U_i(r) \cap U_j(r)} \left|h_i(f_{ij}(x))\right| \quad \text{if} \quad \xi = (f_{ij}) \in Z_b^1(r)\,;$$

$$\|\gamma\|_r = \max_{i} \sup_{x \in U_i(r)} \left|h_i(c_i(x))\right| \quad \text{if} \quad \gamma = (c_i) \in C_b^0(r)\,.$$

With these norms, $Z_b^1(r)$ and $C_b^0(r)$ are Banach spaces. Let $\frac{1}{2} \le \rho < r < 1$. We have: If $\gamma \in C^0(\mathcal{U}(r), E)$ and $\delta\gamma \in Z_b^1(r)$, then $\gamma \in C_b^0(r)$; moreover, there exists a constant $C > 0$ depending only on $\{W_i, z_i, h_i\}$ such that

$$\|\gamma\|_r \le \|\delta\gamma\|_r + C\,\|\gamma\|_\rho\,.$$

In fact, if $\gamma = (c_i)$ and $x_0 \in U_i(r)$; choose j such that $x_0 \in U_j(\rho)$. We have $c_i(x_0) = (c_i - c_j)(x_0) + c_j(x_0)$, and $h_i(c_j(x_0)) = h_i \circ h_j^{-1}\Big(h_j(c_j(x_0))\Big)$, so that

$$\left|h_i(c_j(x_0))\right| \le C\left|h_j(c_j(x_0))\right| \le C\,\|\gamma\|_\rho\,,$$

where C is the supremum of the norms of the matrices $h_i \circ h_j^{-1}(x)$ over $x \in U_i(1) \cap U_j(1)$. Hence

$$\left|h_i(c_i(x_0))\right| \le \|\delta\gamma\|_r + C\,\|\gamma\|_\rho\,.$$

Let $H_b^1(r) = Z_b^1(r)/\delta C_b^0(r)$. Then, the natural map

$$H_b^1(s) \longrightarrow Z^1(\mathcal{U}(s), E)\ /\ \delta C^0(\mathcal{U}(s), E) = H^1(\mathcal{U}(s), E) \simeq H^1(X, E)$$

is an isomorphism for $\frac{1}{2} \le s \le 1$; in fact, the remark above shows that the map is injective. Surjectivity follows from the Leray theorem since the isomorphism $H^1(\mathcal{U}(2), E) \to H^1(\mathcal{U}(s), E)$ factors through $H_b^1(s)$. Also, the restriction map $H_b^1(1) \to H_b^1(s)$ is an isomorphism; in particular, the map $Z_b^1(1) \to H_b^1(r)$ induced by the restriction $Z_b^1(1) \to Z_b^1(r)$ is surjective.

Let N be an integer ≥ 1. Let $\frac{1}{2} \leq \rho < r < 1$ as above, and let $C^0(r,N) = \{\gamma = (c_i) \in C_b^0(r) \mid \operatorname{ord}_{a_i}(c_i) \geq N\}$, where a_i is the point in W_i with $z_i(a_i) = 0$. By Schwarz's lemma (see proof of Theorem 1), we have

$$\|\gamma\|_\rho \leq \left(\frac{\rho}{r}\right)^N \|\gamma\|_r \quad \text{if} \quad \gamma \in C^0(r,N)\,.$$

Thus, if we choose N such that $C\left(\frac{\rho}{r}\right)^N \leq \frac{1}{2}$, we obtain: For $\gamma \in C^0(r,N)$, we have $\|\gamma\|_r \leq \|\delta\gamma\|_r + C\left(\frac{\rho}{r}\right)^N \|\gamma\|_r$, i.e. $\|\gamma\|_r \leq 2\,\|\delta\gamma\|_r$. In particular, $\delta C^0(r,N) \subset Z_b^1(r)$ is a *closed* subspace and the quotient $\mathcal{H} = Z_b^1(r)/\delta C^0(r,N)$ is a Banach space. Moreover, $C_b^0(r)/C^0(r,N)$ is finite dimensional, so that the image of $\delta C_b^0(r)$ in $\mathcal{H}$ has finite dimension and so is *closed* in $\mathcal{H}$ (see the proof of the functional analysis theorem in §8). It follows that $\delta C_b^0(r)$ is closed in $Z_b^1(r)$, and $H_b^1(r) = Z_b^1(r)/\delta C_b^0(r)$ is a (Hausdorff) Banach space.

Now, by Montel's theorem (which asserts that a uniformly bounded sequence of holomorphic functions on an open set Ω in $\mathbb{C}$ has a subsequence converging uniformly on compact subsets of Ω), the restriction map $Z_b^1(1) \to Z_b^1(r)$ $(r < 1)$ is compact [since $U_i(r) \cap U_j(r)$ is relatively compact in $U_i(1) \cap U_j(1)$]. Thus, the induced map $Z_b^1(1) \to H_b^1(r)$ is both compact and surjective. By the open mapping theorem, $H_b^1(r)$ has a relatively compact neighbourhood of 0 (e.g. the image of the open unit ball in $Z_b^1(1)$). Thus, $H_b^1(r) \simeq H^1(X,E)$ is finite dimensional.

Note. The proof of Theorem 2 in an earlier version of these notes used a non-trivial theorem of L. Schwartz on perturbations of surjective linear maps between Banach spaces by compact ones. The arrangement of that proof avoiding Schwartz's theorem as given above was suggested by Madhav Nori.

Theorem 2 is quite powerful. As an immediate application, we shall prove the following theorem.

Theorem 3. *Let X be a compact Riemann surface and $\pi : L \to X$ a holomorphic line bundle. Then L has a meromorphic section which is not holomorphic. In particular: (a) Any line bundle L on X is isomorphic to $L(D)$ for some divisor D on X and (b) there exists a non-constant meromorphic function on X.*

Proof. Let $a \in X$ and let (U,z) be a coordinate neighbourhood of a with $z(a) = 0$; assume also that there is a holomorphic trivialisation $h_U : \pi^{-1}(U) \to U \times \mathbb{C}$.

Let $k \geq 1$ be an integer and s_k be the meromorphic section of L over U for which $h_U(s_k(x)) = \left(x, \frac{1}{(z(x))^k}\right)$, $x \in U - \{a\}$. Consider the covering $\mathcal{U} = \{U, X - \{a\}\}$ and set $f_{12}^{(k)} = s_k \mid U - \{a\}$; set $f_{21}^{(k)} = -f_{12}^{(k)}$ and $f_{ij}^{(k)} = 0$ otherwise $(i,j \in \{1,2\})$. This defines an element $f^{(k)} \in Z^1(\mathcal{U}, L)$. Since $H^1(X,L)$ is finite dimensional and $H^1(\mathcal{U},L) \to$

$H^1(X,L)$ is injective, if $d = \dim_{\mathbb{C}} H^1(X,L)$, there exist constants $c_1, \dots, c_{d+1}$ (not all zero) such that

$$c_1 f^{(1)} + \cdots + c_{d+1} f^{(d+1)} \in B^1(\mathcal{U}, L) ;$$

i.e. there exist holomorphic sections u_1, u_2 of L over U, $X - \{a\}$ respectively such that

$$c_r s_r + \cdots + c_{d+1} s_{d+1} = u_1 - u_2 \quad \text{on} \quad U - \{a\}\, , c_r \neq 0\, .$$

The section $s = u_2$ of L on $X - \{a\}$ is meromorphic (and not holomorphic) on X since $s = -\sum_r^{d+1} c_\nu s_\nu + u_1$ on $U - \{a\}$.

Remark. This argument shows that if $g = \dim H^1(X, \mathcal{O})$ and $a \in X$, there is a (non-constant meromorphic function on X) holomorphic on $X - \{a\}$, with a pole of order $\leq g + 1$ at a.

Theorem 3(b) can be used to prove the following.

Theorem 4. *Let X be a compact Riemann surface and let $\mathcal{M}(X)$ be the field of meromorphic functions on X. Then $\mathcal{M}(X)$ is an algebraic function field in one variable. More precisely, if f is a non-constant meromorphic function on X, $\mathcal{M}(X)$ is a finite algebraic extension of the field $\mathbb{C}(f)$ of rational functions in f.*

Proof. Let f be a non-constant meromorphic function on X. We consider $f : X \to \mathbb{P}^1$ as a holomorphic map into $\mathbb{P}^1$ (the poles map to ∞ in $\mathbb{P}^1 = \mathbb{C} \cup \{\infty\}$). Let $C \subset X$ be the critical points of this map (points where f is not a local homeomorphism) and $B \subset \mathbb{P}^1$ the image of $C : B = f(C)$. B, C are finite and let $A = f^{-1}(C)$. Then $f : X - A \to \mathbb{P}^1 - B$ is a finite covering, say of d sheets.

Let $g \in \mathcal{M}(X)$. We claim that there exist meromorphic functions $a_1, \dots, a_d$ on $\mathbb{P}^1$ such that

$$\big(g(x)\big)^d + a_1\big(f(x)\big)\big(g(x)\big)^{d-1} + \cdots + a_d\big(f(x)\big) = 0\, .$$

To see this, if S is the set of poles of g, we define $a_\nu(z)$ for $z \in \mathbb{P}^1 - B - f(S)$ by $a_\nu(z) =$ ν^{th} elementary symmetric function in $g(x_1), \dots, g(x_d)$, where $\{x_1, \dots, x_d\} = f^{-1}(z)$. Clearly, we have (by definition of elementary symmetric functions),

$$\big(g(x)\big)^d + a_1\big(f(x)\big)\big(g(x)\big)^{d-1} + \cdots + a_d\big(f(x)\big) = 0$$

for $x \in X - A - f^{-1}f(S)$. Thus, we have only to show that the a_ν extend meromorphically to all of $\mathbb{P}^1$.

Let $a \in B \cup f(S)$ and let U be a neighbourhood of a such that the only poles of g on $f^{-1}(U)$ lie in $f^{-1}(a)$, and such that there is a holomorphic function w on U with $w(a) = 0$, $w \not\equiv 0$. Then, there is an integer $N > 0$ such that $(w \circ f)^N g$ is holomorphic on $f^{-1}(U)$. If now W is an open set with $a \in W \subset\subset V$, then $(w \circ f)^N g$ is bounded on $f^{-1}(W)$, so that the νth elementary symmetric function $b_\nu(z)$ of the values of $(w \circ f)^N g$

at $(x_1, \ldots, x_d) = f^{-1}(z)$ $(z \in W - \{a\})$ are bounded, so extend holomorphically to a. But $a_\nu(z) = \frac{b_\nu(z)}{(w(z))^{\nu N}}$, so a_ν has at most a pole at a.

Since any meromorphic function on $\mathbb{P}^1$ is rational this shows that any $g \in \mathcal{M}(X)$ is algebraic over $\mathbb{C}(f)$ of degree $\leq d$.

Choose g_0 such that the degree $\left[\mathbb{C}(f, g_0) : \mathbb{C}(f)\right]$ is maximal. We claim that $\mathbb{C}(f, g_0) = \mathcal{M}(X)$; in fact, if $h \in \mathcal{M}(X)$, $h \notin \mathbb{C}(f, g_0)$, then, since $\mathbb{C}(f)$ has characteristic 0, the field $\mathbb{C}(f)(g_0, h) = \mathbb{C}(f)(g)$ for some $g \in \mathcal{M}(X)$. But then, the degree of g over $\mathbb{C}(f) = \left[\mathbb{C}(f)(g_0, h) : \mathbb{C}(f)\right]$ is greater than $\left[\mathbb{C}(f)(g_0) : \mathbb{C}(f)\right]$, a contradiction.

This proves the theorem.

8. The Dolbeault Isomorphism

In proving Mittag–Leffler's theorem $\left(H^1(U,\mathcal{O}) = 0, U \subset \mathbb{C}\right)$, we reduced the result to solving the equation $\frac{\partial u}{\partial \bar{z}} = f$. The method given there, when formalised, leads to an important interpretation of $H^1(X,E)$ [E being a holomorphic vector bundle on the Riemann surface X] called the Dolbeault isomorphism.

We shall assume that the reader is familiar with the language of exterior algebra and differential forms on a manifold. We shall briefly review the aspects which are of relevance to our discussion.

Let X be a Riemann surface and $T_X^{\mathbb{C}}$ the complex tangent bundle of X (i.e. the bundle of complex valued tangent vectors). Its dual $T_X^{*,\mathbb{C}}$ is the bundle of complex covectors, viz, the complex cotangent bundle, whose C^∞ sections are the smooth differential forms on X. If (U,z) is a coordinate system, and we write $z = x+iy$ (with x, y real-valued), then, a differential form φ on U can be written $\varphi = \varphi_1 dx + \varphi_2 dy$, where $\varphi_1, \varphi_2 \in C^\infty(U)$. If we note that $dz = dx + i\,dy$, $d\bar{z} = dx - i\,dy$, we can also write $\varphi = a\,dz + b\,d\bar{z}$, where $a, b \in C^\infty(U)$.

If now (V,w) is another coordinate system and $f : V \to U$ is a holomorphic map, then $f^*(\varphi) = (a \circ f)f'dw + (b \circ f)\bar{f}'d\bar{w}$ (because of the Cauchy–Riemann equations), so that, if $\varphi = a\,dz + b\,d\bar{z}$, and $b \equiv 0$ on V, then $f^*(\varphi)$ has the property that the coefficient of $d\bar{w}$ is zero; this is thus an intrinsic property of the form. We say that a differential form φ is of type $(1,0)$ [resp. type $(0,1)$] if, for any local coordinate system (U,z), $\varphi|U$ is of the form $\varphi = a\,dz$ [resp. $b\,d\bar{z}$].

For $W \subset X$ open, we set $\mathcal{A}^{1,0}(W)$ [resp. $\mathcal{A}^{0,1}(W)$] = space of 1-forms on W of type $(1,0)$ [resp. $(0,1)$], $\mathcal{A}^{0,0}(W) = C^\infty(W)$.

If $f \in C^\infty(W)$, the exterior derivative df can be written uniquely as $df = \partial f + \bar{\partial} f$ with $\partial f \in \mathcal{A}^{1,0}(W)$, $\bar{\partial} f \in \mathcal{A}^{0,1}(W)$. In local coordinates, we have $\partial f = \frac{\partial f}{\partial z}\,dz$, $\bar{\partial} f = \frac{\partial f}{\partial \bar{z}}\,d\bar{z}$.

If $\alpha \in \mathcal{A}^{1,0}(W)$, we set $\partial\alpha = 0$, $\bar{\partial}\alpha = d\alpha$ (exterior derivative) and similarly, if $\beta \in \mathcal{A}^{0,1}(W)$ $\partial\beta = d\beta$ and $\bar{\partial}\beta = 0$. In local coordinates, if $\alpha = adz$, $\beta = bd\bar{z}$, we have $\bar{\partial}\alpha = \frac{\partial a}{\partial \bar{z}} d\bar{z} \wedge dz = -\frac{\partial a}{\partial \bar{z}} dz \wedge d\bar{z}$ and $\partial\beta = \frac{\partial \beta}{\partial z} dz \wedge d\bar{z}$.

Let $\pi : E \to X$ be a holomorphic vector bundle on the Riemann surface X. If $W \subset X$ is open, we set $C_E^\infty(W)$ = space of C^∞ sections of E over W (i.e. the space of C^∞ maps $s : W \to E$ with $\pi \circ s$ = identity on W). If $E|W$ is trivial (i.e. if there is a holomorphic trivialisation $h : \pi^{-1}(W) \to W \times \mathbb{C}^n$), then the map $H^0(W,E) \otimes_{\mathcal{O}(W)} C^\infty(W) \to C_E^\infty(W)$ [where $H^0(W,E)$ is the space of holomorphic sections of E over W] given by $s \otimes f \mapsto f \cdot s$ is an isomorphism.

Set $\mathcal{A}_E^{0,1}(W) = C_E^\infty(W) \otimes_{C^\infty(W)} \mathcal{A}^{0,1}(W)$, $W \subset X$ open. If $E|W$ is trivial, we have $\mathcal{A}_E^{0,1}(W) = H^0(W,E) \otimes_{\mathcal{O}(W)} \mathcal{A}^{0,1}(W)$.

If W is such that $E|W$ is trivial, there is a unique $\mathcal{O}(W)$-linear map $\bar\partial_{E,W} : C_E^\infty(W) \to \mathcal{A}_E^{0,1}(W)$ induced by the map $1 \otimes \bar\partial : H^0(W,E) \otimes_{\mathcal{O}(W)} C^\infty(W) \to H^0(W,E) \otimes_{\mathcal{O}(W)} \mathcal{A}^{0,1}(W)$ (note that $1 \otimes \bar\partial$ is well-defined since $\bar\partial$ is $\mathcal{O}(W)$-linear). It follows easily that if V is any open set in X, there is a unique $\mathcal{O}(V)$-linear map $\bar\partial_{E,V} : C_E^\infty(V) \to \mathcal{A}_E^{0,1}(V)$ such that, if $U \subset V$ and $E|U$ is trivial, then, for any $s \in C_E^\infty(V)$, $\bar\partial_{E,V}(s)|_U = \bar\partial_{E,U}(s|U)$. We shall denote this map simply by $\bar\partial_E : E \to \mathcal{A}_E^{0,1}$ or just $\bar\partial$.

We can now state the main result of this section.

THE DOLBEAULT ISOMORPHISM. Let $\pi : E \to X$ be a holomorphic vector bundle on the Riemann surface X, and consider the map

$$\bar\partial : C_E^\infty(X) \longrightarrow \mathcal{A}_E^{0,1}(X) .$$

We have: $\ker(\bar\partial) = H^0(X,E)$, the space of holomorphic sections of E over X and $\operatorname{coker}(\bar\partial)$ is naturally isomorphic to $H^1(X,E)$.

Proof. The statement that $\ker(\bar\partial) = H^0(X,E)$ is local. If $U \subset X$ is open, $\pi^{-1}(U) \to U \times \mathbb{C}^n$ is an isomorphism and $s \in C_E^\infty(X)$, then $\bar\partial s|U = 0 \iff \bar\partial f = 0$ where $\big(x, f(x)\big) = h_U\big(s(x)\big)$, $x \in U$. This is the case if and only if f is holomorphic.

To prove the second part, we first prove the following lemma.

Lemma. *Let $\mathbb{E}^\infty$ be the sheaf $\mathbb{E}^\infty(W) = C_E^\infty(W)$, $W \subset X$ open. Then $H^1(X, \mathbb{E}^\infty) = 0$.*

Proof. Let $\mathcal{U} = \{U_i\}_{i \in I}$ be an open covering, and $s_{ij} \in C_E^\infty(U_i \cap U_j)$ be such that $\{s_{ij}\} \in Z^1(\mathcal{U}, \mathbb{E}^\infty)$. Let $\{\alpha_i\}_{i\in I}$ be a partition of unity relative to $\mathcal{U}$, define $s_i \in C_E^\infty(U_i)$ by $s_i = \sum_{j \in I} \alpha_j s_{ij}$ (where $\alpha_j s_{ij}$ is defined by $(\alpha_j s_{ij})(x) = \alpha_j(x) s_{ij}(x)$ if $x \in U_i \cap U_j$, $= 0$ if $x \in U_i - U_i \cap U_j$). Then, as in the proof of the Mittag–Leffler theorem,

$$s_k - s_\ell = \sum_j \alpha_j (s_{kj} - s_{\ell j}) = \sum_j \alpha_j s_{k\ell} = s_{k\ell} \quad \text{on} \quad U_k \cap U_\ell .$$

We define the map $H^1(X,E) \xrightarrow{D} \mathcal{A}_E^{0,1}(X)/\bar\partial C_E^\infty(X)$ as follows: Let $\{s_{ij}\} \in Z^1(U,E)$ $\big(s_{ij} \in H^0(U_i \cap U_j, E)\big)$. Let $\varphi_i \in C_E^\infty(U_i)$ be such that $\varphi_i - \varphi_j = s_{ij}$ on $U_i \cap U_j$. Then $\bar\partial\varphi_i - \bar\partial\varphi_j = 0$ on $U_i \cap U_j$, and so the $\{\bar\partial\varphi_i\}$ define an element of $\mathcal{A}_E^{0,1}(X)$, whose image in the quotient $\mathcal{A}_E^{0,1}(X)/\bar\partial C_E^\infty(X)$ is $D\big(\{s_{ij}\}\big)$.

We check that this is independent of the choices made. First, if $\{V_\alpha\}_{\alpha\in A}$, $\tau : A \to I$ is a refinement of $\{U_i\}$, and $s_{\alpha\beta} = s_{\tau(\alpha)\tau(\beta)}|V_\alpha \cap V_\beta$, we may take $\psi_\alpha = \varphi_{\tau(\alpha)}|V_\alpha$ as the solution of $\psi_\alpha - \psi_\beta = s_{\alpha\beta}$, and we see at once that $\{\bar\partial\psi_\alpha\}$ define the same form as $\{\bar\partial\varphi_i\}$.

If $\varphi_i' \in C_E^\infty(U_i)$ is another solution of $\varphi_i' - \varphi_j' = s_{ij}$, then $\varphi_i' - \varphi_i = \varphi_j' - \varphi_j$ on $U_i \cap U_j$, so defines $\varphi \in C_E^\infty(X)$; clearly if $\omega, \omega' \in \mathcal{A}_E^{0,1}(X)$ are defined by $\omega|U_i$ (resp. $\omega'|U_i) = \bar\partial\varphi_i$ (resp. $\bar\partial\varphi_i'$), we have $\bar\partial\varphi = \omega' - \omega$.

We check that D is injective: given $\{s_{ij}\} \in Z^1(\mathcal{U}, E)$ and $\varphi_i \in C_E^\infty(U_i)$ with $\varphi_i - \varphi_j = s_{ij}$ on $U_i \cap U_j$, and if $\varphi \in C_E^\infty(X)$ is such that $\bar\partial\varphi = \bar\partial\varphi_i$ on U_i, then, we have

$$(\varphi_i - \varphi) - (\varphi_j - \varphi) = s_{ij} \quad \text{on} \quad U_i \cap U_j \,, \bar\partial(\varphi_i - \varphi) = 0 \,, \text{i.e. } \varphi_i - \varphi \in H^0(U_i, E) \,.$$

Finally, D is surjective. In fact, given $\omega \in \mathcal{A}_E^{0,1}(X)$, we can find an open covering $\mathcal{U} = \{U_i\}_{i\in I}$ and $\varphi_i \in C_E^\infty(U_i)$ with $\bar\partial\varphi_i = \omega|U_i$ [because, if $\Omega \subset \mathbb{C}$ is open and if $f \in C^\infty(\Omega)$ then, by case 1 of proposition 3 in §5, if $K \subset \Omega$ is compact, then we can find $u \in C^\infty(\Omega)$ with $\frac{\partial u}{\partial \bar z} = f$ on K]. Moreover, if $s_{ij} = \varphi_i - \varphi_j|U_i \cap U_j$, then $\bar\partial s_{ij} = 0$ on $U_i \cap U_j$, so that $s_{ij} \in H^0(U_i \cap U_j, E)$. Clearly $\{s_{ij}\} \in Z^1(\mathcal{U}, E)$. By construction, $D(\{s_{ij}\}) =$ class of ω.

Note. We could simply have used the short exact sequence of sheaves $0 \to \mathbb{E} \to \mathbb{E}^\infty \to \mathcal{A}_E^{0,1} \to 0$ together with the lemma $\big(H^1(X, \mathbb{E}^\infty) = 0\big)$; the exact cohomology sequence then implies the theorem. However, the actual construction of the map D given above is often useful.

The Dolbeault isomorphism and the finiteness theorem of §7 have an important consequence.

First, we define topologies on $C_E^\infty(X)$ and $\mathcal{A}_E^{0,1}(X)$. Let $a \in X$, and let U be a neighbourhood of a carrying a coordinate z ($z : U \to z(U) \subset \mathbb{C}$ being an isomorphism) and such that $E|U$ is trivial. If $\varphi \in \mathcal{A}_E^{0,1}(X)$, then $\varphi|U$ can be identified with an expression.

$$(\varphi_1, \ldots, \varphi_n)\, d\bar z \,, \ \varphi_1, \ldots, \varphi_n \in C^\infty(U) \,.$$

Let X be a *compact* Riemann surface. We introduce a topology of complete metric space (even a Fréchet space) on $\mathcal{A}_E^{0,1}(X)$ by the following requirement:

A sequence $\varphi^{(\nu)} \in \mathcal{A}_E^{0,1}(X)$ converges if and only if for any U as above, the corresponding sequence $(\varphi_1^{(\nu)}, \ldots, \varphi_n^{(\nu)})$

$$\left[(\varphi_1^{(\nu)}, \ldots, \varphi_n^{(\nu)})\, d\bar z = \varphi^{(\nu)}|U\right]$$

converges in $\prod_{n \text{ copies}} C^\infty(U)$ [i.e., for any differentiation D^k of order k $\big(D^k = \frac{\partial^k}{\partial x^\ell \partial y^m}, \ell + m = k\big)$, the sequence $\{D^k \varphi_j^{(\nu)}\}$ converges uniformly on compact subsets of U].

We introduce, in the same way, a topology on $C_E^\infty(X)$. These are called the C^∞ (or Schwartz) topologies on $\mathcal{A}_E^{0,1}, C_E^\infty$.

Theorem. *$\bar\partial C_E^\infty(X)$ is a closed subspace of $\mathcal{A}_E^{0,1}(X)$.*

Since $\bar{\partial}C_E^\infty(X)$ has finite codimension in $\mathcal{A}_E^{0,1}(X)$ [by the finiteness of $\dim H^1(X,E)$ and the Dolbeault isomorphism], this is a consequence of the following standard theorem in functional analysis.

Let V, W be Fréchet spaces and $u : V \to W$ a continuous linear map.
If $\dim_{\mathbb{C}}\big(W/u(V)\big) < \infty$, then $u(V)$ is closed in W. (Note. It is not true that a finite codimensional subspace of a Fréchet space is closed.)

We outline the proof. First, since $\ker(u)$ is closed, $V/\ker(u)$ is again Fréchet, so we may assume that u is injective. Let $W_0 \subset W$ be a finite dimensional space such that the projection $W_0 \to W/u(V)$ is an algebraic isomorphism. Now, W_0 is closed in W. [If $w_1, \ldots, w_k$ is a basis of W_0, then the map $\mathbb{C}^k \to W_0$, $(x_1, \ldots, x_k) \mapsto \sum x_i w_i$ is continuous and bijective; it is easily seen that it is a homeomorphism because the image of $\{\sum |x_i|^2 = 1\}$ is a compact, hence closed subset of W_0 so that there is a neighbourhood of 0 disjoint from this image. It follows that W_0 is complete in the induced topology from W, hence closed.] The map $W_0 \oplus V \longrightarrow W$, $(w,v) \mapsto w + u(v)$ is a continuous bijection, hence a homeomorphism (open mapping theorem). Since V is closed in $W_0 \oplus V$, its image is closed in W.

We shall end this section by describing the canonical bundle of a Riemann surface X.

Let $W \subset X$ be open. A holomorphic 1-form on W is a 1-form ω of type $(1,0)$ such that $\bar{\partial}\omega = 0$ (i.e. such that $d\omega = 0$). If (U, z) is a coordinate system, and $\omega|W \cap U = f dz$, then, if ω is holomorphic, so is f. Meromorphic 1-forms are forms defined outside a discrete set in W which, for any coordinate system (U,z) with $U \subset W$ are of the form $f dz$ with f meromorphic on U.

Let $\Omega = \Omega_X (= \Omega_X^1)$ be the sheaf $U \mapsto \Omega_X(U) = \{\text{space of holomorphic 1-forms on } U\}$; it is called the sheaf of holomorphic 1-forms. There is a line bundle $K = K_X$ on X such that, for any open set $U \subset X$, we have $H^0(U, K_X) = \Omega_X(U)$.

This line bundle can be described intrinsically by analyzing the complex cotangent bundle $T_X^{*,\mathbb{C}}$ and decomposing complex covectors into those of type $(1,0)$ and those of type $(0,1)$. Here we shall simply give one system of transition functions defining it.

Let $\{(U_i, z_i)\}$ be a covering of X by coordinate neighbourhoods. Clearly, there are functions $g_{ij} \in \mathcal{O}(U_i \cap U_j)$, holomorphic and nowhere zero on $U_i \cap U_j$ such that

$$dz_j = g_{ij} \cdot dz_i \quad \text{on} \quad U_i \cap U_j \,.$$

Let K_X be the line bundle defined by the transition function $\{g_{ij}\}$. If $W \subset X$ is open and $s \in H^0(W, K_X)$, then s is given by a family $\{f_i\}$, $f_i \in \mathcal{O}(W \cap U_i)$ such that

$$f_i = g_{ij} f_j \quad \text{on} \quad W \cap U_i \cap U_j \,.$$

Consider the holomorphic 1-forms $\omega_i = f_i\, dz_i$ on $W \cap U_i$; we have, on $W \cap U_i \cap U_j$, $\omega_i = f_i\, dz_i = g_{ij} f_j\, dz_i = f_j\, dz_j = \omega_j$, so that they define $\omega \in \Omega_X(U)$. The map $s \mapsto \omega$ is clearly an isomorphism.

The line bundle $K = K_X$ is called *the canonical (line) bundle* of X. Under the correspondence described above, meromorphic sections of K_X correspond to meromorphic 1-forms. We may thus speak of the divisor of a meromorphic 1-form ω; such divisors are called canonical divisors.

Note also that the above construction, applied to C^∞ sections of K_X over an open set $W \subset X$ gives an isomorphism of $C^\infty_{K_X}(W)$ with $\mathcal{A}^{1,0}(W)$.

9. Weyl's Lemma and the Serre Duality Theorem

What is usually called Weyl's lemma is a regularity theorem for the Laplacian rather the operator $\bar{\partial}$.

Let X be a compact Riemann surface and $\pi : E \to X$ a holomorphic vector bundle on X. Let $\pi^* : E^* \to X$ be the dual vector bundle and let K_X be the canonical line bundle on X. We define a bilinear form $\langle\ ,\ \rangle$ $H^0(X, E^* \otimes K_X) \times \mathcal{A}_E^{0,1}(X) \to \mathbb{C}$ as follows.

Let $s \in H^0(X, E^* \otimes K_X)$, $\varphi \in \mathcal{A}_E^{0,1}(X)$, and let (U, z) be a coordinate neighbourhood on X. Let ω be a holomorphic 1-form on U without zeros (i.e. if $\omega = f\,dz$, then $f \neq 0$ at every point of U). Then s, φ can be written uniquely

$$s|U = \lambda \otimes \omega\ , \quad \varphi|U = \alpha \otimes \bar{\omega} \quad \text{on} \quad U\ ,$$

where $\lambda \in H^0(U, E^*)$, $\alpha \in C_E^\infty(U)$; we define a C^∞ function (λ, α) on U by $(\lambda, \alpha)(x) = \lambda(x)(\alpha(x))$, $x \in U$ [$\lambda(x)$ is linear form on $E_x = \pi^{-1}(x)$ and $\alpha(x) \in E_x$], and the 2-form

$$(s|U, \varphi|U) = (\lambda, \alpha)\omega \wedge \bar{\omega}$$

is independent of the choice of ω on U [if $\omega' = f\omega$ with f holomorphic and non-zero on U, then $(\frac{1}{f}\lambda) \otimes \omega' = s|U$, and $\varphi|U = (\frac{1}{\bar{f}}\alpha) \otimes \bar{\omega}'$, so that $(\frac{1}{f}\lambda, \frac{1}{\bar{f}}\alpha)\omega' \wedge \bar{\omega}' = \frac{1}{|f|^2}(\lambda, \alpha) f\bar{f}\omega \wedge \bar{\omega} = (\lambda, \alpha)\omega \wedge \bar{\omega}$]. This defines a 2-form (s, φ) on X, and we set

$$\langle s, \varphi \rangle = \int_X (s, \varphi)\ .$$

Note that if $W \subset X$ is open and $\operatorname{supp}(\varphi) \subset W$, then $\langle s, \varphi \rangle$ is defined for all $s \in H^0(W, E^* \otimes K_X)$.

Let (U, z) be a coordinate neighbourhood such that there is a trivialisation $h_U : \pi^{-1}(U) \to U \times \mathbb{C}^n$ and let $h_U^* : \pi^{*-1}(U) \to U \times \mathbb{C}^n$ be the corresponding trivialisation of the dual (= inverse of the transpose). Then, if $v \in E_x$ and $\lambda \in E_x^*$, $x \in U$, and $h_U(v) = (x, v_1, \dots, v_n)$, $h_U^*(\lambda) = (x, \lambda_1, \dots, \lambda_n)$, we have $\lambda(v) = \sum_1^n \lambda_k v_k$. If $s \in H^0(X, E^* \otimes K_X)$ and $\varphi \in \mathcal{A}_E^{0,1}(X)$ and if $\operatorname{supp}(\varphi) \subset U$, we can write

$$\begin{aligned} s|U = \lambda \otimes dz \quad &\text{with} \quad h_U^*(\lambda(x)) = (x, \lambda_1(x), \dots, \lambda_n(x)) \\ \varphi = \alpha \otimes d\bar{z} \quad &\text{with} \quad h_U(\alpha(x)) = (x, \alpha_1(x), \dots, \alpha_n(x)) \end{aligned}$$

and we have

$$\langle s, \varphi \rangle = \int_U \Big(\sum_{k=1}^n \lambda_k(z)\alpha_k(z)\Big)\, dz \wedge d\bar{z}\ .$$

We make the following remark. If $f \in C^\infty_E(X)$, and $s \in H^0(X, E^* \otimes K_X)$, then $\langle s, \bar\partial f\rangle = 0$. To see this, f can be written (partition of unity) as $\sum_{\nu=1}^p f_\nu$, where each f_ν has its support in a coordinate neighbourhood (U_ν, z_ν) on which $E|U_\nu$ is trivial. We therefore suppose that $p = 1$ that $\operatorname{supp}(f) \subset U$, with (U, z) as above. If $h_U\big(f(x)\big) = \big(x, f_1(x), \dots, f_n(x)\big)$, then $\bar\partial f = \alpha \otimes d\bar z$ with $h_U\big(\alpha(z)\big) = \Big(z, \frac{\partial f_1}{\partial \bar z}, \dots, \frac{\partial f_n}{\partial \bar z}\Big)$; hence, if $h_U^*\big(s(z)\big) = \big(z, \lambda_1(z), \dots, \lambda_n(z)\big)$ on U, we have

$$\langle s, \varphi\rangle = \int_U \sum \lambda_k(z) \frac{\partial f_k}{\partial \bar z}\, dz \wedge d\bar z = -\int_U \sum \frac{\partial \lambda_k}{\partial \bar z} f_k dz \wedge d\bar z = 0$$

[by integration by parts, since $\operatorname{supp}(f_k) \subset U$].

WEYL'S LEMMA: **The regularity theorem for $\bar\partial$.**

Let X be a compact Riemann surface, let $\pi : E \to X$ be a holomorphic vector bundle on X. We equip $\mathcal{A}_E^{0,1}(X)$ with the C^∞ topology (described in §8), viz: the topology of convergence of all derivatives on compact subsets of coordinate neighbourhoods in X.

Suppose that $F : \mathcal{A}_E^{0,1}(X) \to \mathbb{C}$ is a continuous linear form such that $F|\bar\partial C^\infty_E(X) = 0$. Then, there exists a unique $s \in H^0(X, E^* \otimes K_X)$ such that

$$F(\varphi) = \langle s, \varphi\rangle \ \forall \varphi \in \mathcal{A}_E^{0,1}(X)\,.$$

Remarks. 1. If E is the trivial bundle of rank 1, and X is an open set in $\mathbb{C}$, we would be dealing with linear forms on $C_0^\infty(X)$ (functions with compact support); see remark 2 relative to the compactness of X. In this context, they are called distributions. Derivatives of distributions are so defined that the condition $F|\bar\partial C_0^\infty(X) = 0$ simply means that $\frac{\partial F}{\partial \bar z} = 0$ in the sense of distributions. Thus the Weyl lemma asserts that distributions F with $\frac{\partial F}{\partial \bar z} = 0$ are given by holomorphic functions.

2. The compactness of X is irrelevant; one should deal simply with the space of $(0, 1)$ forms with values in E which have compact support. To treat the general case, one must define the C^∞ or Schwartz topology on this space of sections. This involves inductive limit topologies which we have preferred to avoid. Some care has to be paid to supports, and we shall have to do this in our proof, although this is, for what we need, rather simple.

Proof of Weyl's lemma. We begin with the following remark. If $W \subset X$ is open, $\sigma \in H^0(W, E^* \otimes K)$ and $\langle \sigma, \varphi\rangle = 0 \ \ \forall \varphi \in \mathcal{A}_E^{0,1}(X)$ with $\operatorname{supp}(\varphi) \subset W$, then $\sigma = 0$.

Hence, it suffices to prove the following:

Let (U, z) be a coordinate neighbourhood such that there is a holomorphic trivialisation $h_U : \pi^{-1}(U) \to U \times \mathbb{C}^n$. Then, there exists $s \in H^0(U, E^* \otimes K_X)$ such that $F(\varphi) = \langle s, \varphi\rangle \ \forall \varphi \in \mathcal{A}_E^{0,1}(X)$ with $\operatorname{supp}(\varphi) \subset U$.

This is equivalent to showing that there exist holomorphic functions $\lambda_1,\ldots,\lambda_n$ on U such that

$$F(\varphi)=\int_U\sum_1^n\lambda_k(z)\alpha_k(z)dz\wedge d\bar z\,,\quad \Big(\varphi=\alpha\otimes d\bar z, h_U\big(\alpha(z)\big)=\big(z,\alpha_1(z),\ldots,\alpha_n(z)\big)\Big)$$

$\forall C^\infty$ functions $\alpha_1,\ldots,\alpha_n$ on U with $\mathrm{supp}(\alpha_j)$ compact in U, (i.e. $\forall\alpha_j\in C_0^\infty(U)$). If we set for $\alpha_j\in C_0^\infty(U)$, $G(\alpha_1,\ldots,\alpha_n)=F(\alpha\otimes d\bar z)$, $h_U\big(\alpha(z)\big)=\big(z,\alpha_1(z),\ldots,\alpha_n(z)\big)$ the condition $F|\bar\partial C_E^\infty(X)=0$ implies that $G(\frac{\partial\beta_1}{\partial\bar z},\ldots,\frac{\partial\beta_n}{\partial\bar z})=0$ if $\beta_j\in C_0^\infty(U)$. It is sufficient to show that $\forall k$, $1\le k\le n$, there exists $\lambda_k\in\mathcal O(U)$ such that

$$G(0,\ldots,\alpha_k,\ldots 0)=\int_U\lambda_k\alpha_k dz\wedge d\bar z\quad\forall\alpha_k\in C_0^\infty(U)\,.$$

Thus, Weyl's lemma is a consequence of the following result, usually known as the regularity theorem for $\frac{\partial}{\partial\bar z}$.

Theorem. *Let U be open in $\mathbb C$. Let $T:C_0^\infty(U)\to\mathbb C$ be a $\mathbb C$-linear map with the following two properties:*

(1) *If $\alpha^{(\nu)}\in C_0^\infty(U)$ is a sequence with supports in a fixed compact set in U and converging in the C^∞ topology to $\alpha\in C_0^\infty(U)$, then $T(\alpha^{(\nu)})\to T(\alpha)$.*

(2) *$T(\frac{\partial\beta}{\partial\bar z})=0$ if $\beta\in C_0^\infty(U)$.*

Then, there exists $\lambda\in\mathcal O(U)$ such that $T(\alpha)=\int_U\lambda\alpha\,dz\wedge d\bar z$ $\forall\alpha\in C_0^\infty(U)$.

Proof of the theorem. We may assume that U is a bounded open set in $\mathbb C$. Let $\varepsilon>0$, and define $U_\varepsilon=\{z\in U|$ distance of z to $\mathbb C-U$ is $>\varepsilon\}$.

Let $\varphi\in C_0^\infty(\mathbb C)$ be such that $\varphi(z)=1$ for $|z|<\frac12\varepsilon$, $\varphi(z)=0$ for $|z|\ge\varepsilon$, $0\le\varphi\le1$ everywhere.

If $\alpha\in C_0^\infty(U_\varepsilon)$, define $\tilde\alpha\in C_0^\infty(U)$ by

$$\tilde\alpha(z)=\frac{1}{2\pi i}\int_{\mathbb C}\alpha(z+w)\frac{\varphi(w)}{w}dw\wedge d\bar w\,.$$

We claim that

$$(*)\qquad \alpha(z)=\frac{\partial\tilde\alpha}{\partial\bar z}(z)+\frac{1}{2\pi i}\int_{\mathbb C}\alpha(z+w)\rho(w)dw\wedge d\bar w\,,\ z\in U\,,$$

where $\rho(w)=\frac{\partial}{\partial\bar w}(\frac{\varphi(w)}{w})$, $w\ne0$, $\rho(0)=0$; ρ is C^∞ and has support in the disc $|w|\le\varepsilon$ (in fact in the annulus $\frac12\varepsilon\le|w|\le\varepsilon$).

To prove $(*)$, fix $z \in U$ and let Δ_δ be a small disc of radius $\delta > 0$ around z. Then

$$\frac{\partial \tilde{\alpha}}{\partial \bar{z}} = \frac{1}{2\pi i}\int_{\mathbb{C}} \frac{\partial \alpha}{\partial \bar{z}}(z+w)\frac{\varphi(w)}{w} dw \wedge d\bar{w} = \lim_{\delta \to 0} \frac{1}{2\pi i} \int_{\mathbb{C}-\Delta_\delta} \frac{\partial \alpha}{\partial \bar{w}}(z+w)\frac{\varphi(w)}{w} dw \wedge d\bar{w} ;$$

the latter integral, for fixed $\delta > 0$, is

$$-\int_{\mathbb{C}-\Delta_\delta} d\Big(\alpha(z+w)\frac{\varphi(w)}{w}\, dw\Big) - \int_{\mathbb{C}-\Delta_\delta} \alpha(z+w)\frac{\partial}{\partial \bar{w}}\Big(\frac{\varphi(w)}{w}\Big)\, dw \wedge d\bar{w} ;$$

the first term $= \int_{|w|=\delta} \frac{\alpha(z+w)\varphi(w)}{w}\, dw$, which converges to $2\pi i\, \alpha(z)\varphi(0) = \alpha(z)$ as $\delta \to 0$. Thus

$$\frac{\partial \tilde{\alpha}}{\partial \bar{z}} = \alpha(z) - \frac{1}{2\pi i}\int_{\mathbb{C}} \alpha(z+w)\rho(w) dw \wedge d\bar{w} ,$$

which proves $(*)$. This can be written

$$\alpha(z) = \frac{\partial \tilde{\alpha}}{\partial \bar{z}} + \frac{1}{2\pi i}\int_U \alpha(w)\rho(w-z) dw \wedge d\bar{w} .$$

Now, if we approximate this integral by Riemann sums, these sums converge to the integral in the C^∞ topology on U (since $\rho \in C^\infty$) and have support in a fixed compact set in U since $\alpha \in C_0^\infty(U_\varepsilon)$ and $\rho(w-z) = 0$ if $|w-z| \geq \varepsilon$. Hence, by the continuity hypothesis (1), we have

$$T(\alpha) = T\Big(\frac{\partial \tilde{\alpha}}{\partial \bar{z}}\Big) + \frac{1}{2\pi i}\int_U \alpha(w)\lambda(w) dw \wedge d\bar{w} ,$$

where, for $w \in U_\varepsilon$, $\lambda(w) = T\big(z \mapsto \rho(w-z)\big)$. Moreover, $\lambda \in C^\infty(U_\varepsilon)$ (since, e.g. if $h \neq 0$ is real and $\to 0$, $\frac{\lambda(w+h)-\lambda(w)}{h} = T\big(\frac{\rho(w+h-z)-\rho(w-z)}{h}\big)$ and the term under T converges in the C^∞ topology when $h \to 0$; we can iterate this argument). Moreover, $T\big(\frac{\partial \tilde{\alpha}}{\partial \bar{z}}\big) = 0$. Thus, we have shown that there is $\lambda \in C^\infty(U_\varepsilon)$ such that

$$T(\alpha) = \frac{1}{2\pi i}\int_U \alpha(w)\lambda(w) dw \wedge d\bar{w} \quad \forall \alpha \in C_0^\infty(U_\varepsilon) .$$

If $\beta \in C_0^\infty(U_\varepsilon)$, we have

$$0 = T\Big(\frac{\partial \beta}{\partial \bar{z}}\Big) = \frac{1}{2\pi i}\int_U \frac{\partial \beta}{\partial \bar{w}}\lambda(w) dw \wedge d\bar{w} = -\frac{1}{2\pi i}\int_U \beta \frac{\partial \lambda}{\partial \bar{w}}\, dw \wedge d\bar{w} ,$$

since $\beta \in C_0^\infty(U_\varepsilon)$ is arbitrary, it follows that $\frac{\partial \lambda}{\partial \bar{w}} = 0$ on U_ε, i.e. that $\lambda \in \mathcal{O}(U_\varepsilon)$. The theorem is proved.

Let X be a compact Riemann surface, $\pi : E \to X$ a holomorphic vector bundle on X. We define a bilinear form $\langle\ ,\ \rangle_E : H^0(X, E^* \otimes K_X) \times H^1(X, E) \to \mathbb{C}$ as follows.

Consider the bilinear form $\langle\ ,\ \rangle : H^0(X, E^* \otimes K_X) \times \mathcal{A}_E^{0,1}(X) \to \mathbb{C}$ defined above. We have seen that $\langle s, \bar\partial f\rangle = 0$ if $f \in C_E^\infty(X)$, $s \in H^0(X, E^* \otimes K_X)$, hence this form induces a bilinear form $H^0(X, E^* \otimes K_X) \times [\mathcal{A}_E^{0,1} \,/\, \bar\partial C_E^\infty(X)] \to \mathbb{C}$, which we denote again by $\langle\ ,\ \rangle$. If we denote by D the Dolbeault isomorphism $D : H^1(X, E) \to \mathcal{A}_E^{0,1}(X) \,/\, \bar\partial C_E^\infty(X)$, we define $\langle\ ,\ \rangle_E$ by

$$\langle s, \xi\rangle_E = \langle s, D(\xi)\rangle, \ s \in H^0(X, E^* \otimes K_X), \ \xi \in H^1(X, E) .$$

This bilinear form induces a map

$$H^0(X, E^* \otimes K_X) \xrightarrow{\Delta_E} H^1(X, E)^*$$

[V^* is the dual of the vector space V], viz

$$\Delta_E(s) \quad \text{is the linear map} \quad \xi \longmapsto \langle s, \xi\rangle_E \quad \text{from} \quad H^1(X, E) \text{ to } \mathbb{C} .$$

SERRE'S DUALITY THEOREM. The map $\Delta_E : H^0(X, E^* \otimes K_X) \to H^1(X, E)^*$ is an isomorphism for any holomorphic vector bundle on the compact Riemann surface X.

Proof. Let $\ell : \mathcal{A}_E^{0,1}(X) \,/\, \bar\partial C_E^\infty(X) \to \mathbb{C}$ be any $\mathbb{C}$-linear map. We have seen that $\bar\partial C_E^\infty(X)$ is closed in $\mathcal{A}_E^{0,1}(X)$ (for the C^∞ topology) and is of finite codimension. Since any linear form on a finite dimensional (Hausdorff) topological vector space is continuous, the $\mathbb{C}$-linear map $F : \mathcal{A}_E^{0,1}(X) \to \mathbb{C}$ given by $F(\varphi) = \ell(\{\varphi\})$, $\{\varphi\}$ = image of φ in $\mathcal{A}_E^{0,1}(X) \,/\, \bar\partial C_E^\infty(X)$, is continuous for the C^∞ topology, and is zero on $\bar\partial C_E^\infty(X)$. By Weyl's lemma, there is $s \in H^0(X, E^* \otimes K_X)$ such that

$$\ell(\{\varphi\}) = F(\varphi) = \langle s, \varphi\rangle = \langle s, \{\varphi\}\rangle \quad \forall \varphi \in \mathcal{A}_E^{0,1}(X) .$$

Since D is an isomorphism this proves that Δ_E *is surjective.*

To prove that Δ_E is injective, given $s \in H^0(X, E^* \otimes K_X)$, we must show that if $\langle s, \varphi\rangle = 0 \ \ \forall \varphi \in \mathcal{A}_E^{0,1}(X)$, then $s \equiv 0$. This is immediate (the remark starting the proof of Weyl's lemma).

In view of the fact that $H^0(X, E)$ and $H^1(X, E)$ are finite dimensional, the Serre duality theorem can be stated as follows:

Alternate form of the duality theorem. The bilinear form

$$\langle\ ,\ \rangle_E : H^0(X, E^* \otimes K_X) \times H^1(X, E) \to \mathbb{C}$$

is non-degenerate (or perfect).

If F is a holomorphic vector bundle, and we take $E = F^* \otimes K_X$, then $E^* \otimes K_X = F \otimes K_X^* \otimes K_X = F$ (since, for any line bundle L, $L^* \otimes L$ is canonically trivial). Hence $H^0(X, F)$ is isomorphic to the dual of $H^1(X, F^* \otimes K_X)$ for any vector bundle F on X.

We shall end this section by singling out the special case of these results when E is the trivial bundle of rank 1 [so that $\mathcal{A}_E^{0,1} = \mathcal{A}^{0,1}(X)$ and $H^1(X, E) = H^1(X, \mathcal{O})$].

Proposition. *Let φ be a form of type $(0,1)$ on the compact Riemann surface X. Then, $\exists f \in C^\infty(X)$ such that $\bar{\partial} f = \varphi$ if and only if, for every holomorphic 1-form ω on X, we have*

$$\int_X \omega \wedge \varphi = 0 .$$

10. The Riemann–Roch Theorem and some Applications

Throughout this section, X will be a compact Riemann surface.

We begin with some definitions.

Let $D = \sum_{i=1}^{r} n_i P_i$ be a divisor on X. The integer $d = \sum_{i=1}^{r} n_i$ is called the degree of D [and written $\deg D$].

Let ω be a meromorphic 1-form, $\omega \not\equiv 0$. If $a \in X$, we denote by $\mathrm{res}_a(\omega)$ (the residue at a of ω) the following: If (U, z) is a local coordinate with $z(a) = 0$, and $\omega = f\,dz$, then $\mathrm{res}_a(\omega) =$ residue of f at $a =$ the coefficient of $\frac{1}{z}$ in the Laurent expansion $\sum_{-N}^{\infty} c_\nu z^\nu$ of f. It is independent of the coordinate system. In fact, if γ is a piecewise differentiable curve in $U - \{a\}$ whose winding number ($=$ index) at a is $+1$, then $\mathrm{res}_a(\omega) = \frac{1}{2\pi i}\int_\gamma \omega$.

Lemma 1. *Let ω be a meromorphic 1-form, $\omega \not\equiv 0$, on X. We have $\sum_{a\in X} \mathrm{res}_a(\omega) = 0$.*

Proof. Let $a_1, \ldots, a_m$ be the poles of ω. Choose coordinate neighbourhoods (U_j, z_j) about a_j $\big(z_j(a_j) = 0\big)$ and let $\Delta_j = \{x \in U \mid |z_j(x)| < \varepsilon\}$ ($\varepsilon > 0$ small). Let $U = X - \bigcup_{j=1}^{m} \Delta_j$. By Stokes' theorem, we have

$$\int_U d\omega = -\sum_j \int_{\partial\Delta_j} \omega = -2\pi i \sum_j \mathrm{res}_{a_j}(\omega)\ .$$

But, a holomorphic 1-form on a Riemann surface is closed: if $\omega = f\,dz$, $d\omega = \frac{\partial f}{\partial z}\,dz \wedge dz + \frac{\partial f}{\partial \bar z}\,d\bar z \wedge dz = 0$.

Corollary 1. Let f be a meromorphic function, $f \not\equiv 0$, on X. Then the degree of the divisor (f) of f is zero.

In fact, $\deg(f) = \sum_{a\in X} \mathrm{res}_a(\omega)$ where $\omega = \frac{1}{f}\,df$.

Corollary 2. If D_1 and D_2 are divisors on X and if D_1 is linearly equivalent to D_2, then $\deg(D_1) = \deg(D_2)$.

In fact, if $D_1 - D_2 = (f)$, we have $\deg(D_1) - \deg(D_2) = \deg(f)$.

Let D be a divisor on X; we introduced the sheaf $\mathcal{O}_D$: $U \mapsto \mathcal{O}_D(U) = \{f$ meromorphic on $U \mid (f) \geq -D$ on $U\}$. $\mathcal{O}_D$ is isomorphic to the sheaf $U \mapsto H^0\big(U, L(D)\big)$ of holomorphic sections of the line bundle defined by D. An element f in the stalk $\mathcal{O}_{D,a}$ is given by a (convergent) Laurent series

$$f = \sum_{n \geq -D(a)} c_n z^n \qquad \big((U, z) \text{ a local coordinate with } z(a) = 0\big)\ .$$

Let D_1, D_2 be divisors with $D_1 \leq D_2$. Clearly $\mathcal{O}_{D_1}(U) \subset \mathcal{O}_{D_2}(U)$ for all U open in X and we obtain an injective morphism of sheaves $\mathcal{O}_{D_1} \to \mathcal{O}_{D_2}$. [In terms of the line bundles $L(D_1), L(D_2)$ and the isomorphism $L(D_1) \otimes L(D_2 - D_1) \simeq L(D_2)$, the map is given by $f \in H^0(U, L(D_1))$, $f \mapsto f \otimes s_{D_2-D_1}$, where $s_{D_2-D_1}$ is the standard section of $L(D_2 - D_1)$ with $(s_{D_2-D_1}) = D_2 - D_1$.] Let $S_{D_1}^{D_2}$ be the sheaf associated to the presheaf $U \mapsto \mathcal{O}_{D_2}(U)/\mathcal{O}_{D_1}(U)$. We have $\left(S_{D_1}^{D_2}\right)_a = 0$ if $D_1(a) = D_2(a)$ (in particular if a is not in $\operatorname{supp}(D_1) \cup \operatorname{supp}(D_2)$). If $D_1(a) < D_2(a)$, then $\left(S_{D_1}^{D_2}\right)_a$ is a finite dimensional vector space, isomorphic to the vector space $\left\{\sum_{-D_2(a) \leq n < -D_1(a)} c_n z^n \mid c_n \in \mathbb{C}, z \text{ an indeterminate}\right\}$.

Lemma 2. *If $D_1 \leq D_2$ are divisors on X, we have*

(1) $\dim H^0(X, S_{D_1}^{D_2}) = \deg D_2 - \deg D_1$,

(2) $H^1(X, S_{D_1}^{D_2}) = 0$.

Proof. (1) $S_{D_1}^{D_2}$ is a sheaf supported at finitely many points ($\in \operatorname{supp}(D_1) \cup \operatorname{supp}(D_2)$). We therefore have

$$\begin{aligned}\dim H^0(X, S_{D_1}^{D_2}) &= \dim \prod_{a \in X} \left(\mathcal{O}_{D_2,a} \,/\, \mathcal{O}_{D_1,a}\right) \\ &= \sum_{a \in X} (D_2(a) - D_1(a)) = \deg D_2 - \deg D_1 \,.\end{aligned}$$

(2) Given any open covering $\mathcal{U}$, there is a refinement $\mathcal{V} = \{V_\alpha\}$ such that, if $\alpha \neq \beta$, we have $V_\alpha \cap V_\beta \cap (\operatorname{supp}(D_1) \cup \operatorname{supp}(D_2)) = \emptyset$. Clearly then $Z^1(\mathcal{V}, S_{D_1}^{D_2}) = 0$, and (2) follows.

Some notation. If D is a divisor, we set $h^i(D) = \dim_{\mathbb{C}} H^i(X, \mathcal{O}_D) = \dim_{\mathbb{C}} H^i(X, L(D))$ $(i = 0, 1)$, and $\chi(D) = h^0(D) - h^1(D)$.

Lemma 3. *If D_1, D_2 are divisors, we have*

$$\chi(D_2) - \deg D_2 = \chi(D_1) - \deg D_1 \,.$$

Proof. Let D_0 be a divisor with $D_0 \leq D_\nu$, $\nu = 1, 2$. We have a short exact sequence of sheaves

$$0 \longrightarrow \mathcal{O}_{D_0} \longrightarrow \mathcal{O}_{D_\nu} \longrightarrow S_{D_0}^{D_\nu} \longrightarrow 0$$

and hence a cohomology sequence

$$\begin{aligned}0 \longrightarrow H^0(X, \mathcal{O}_{D_0}) &\longrightarrow H^0(X, \mathcal{O}_{D_\nu}) \longrightarrow H^0(X, S_{D_0}^{D_\nu}) \longrightarrow \\ \longrightarrow H^1(X, \mathcal{O}_{D_0}) &\longrightarrow H^1(X, \mathcal{O}_{D_\nu}) \longrightarrow 0\end{aligned}$$

($H^1(X, S_{D_0}^{D_\nu}) = 0$ by Lemma 2). If V is the image of $H^0(X, \mathcal{O}_{D_\nu})$ in $H^0(X, S_{D_0}^{D_\nu})$, we have two exact sequences

$$0 \longrightarrow H^0(X, \mathcal{O}_{D_0}) \longrightarrow H^0(X, \mathcal{O}_{D_\nu}) \longrightarrow V \longrightarrow 0$$

$$0 \longrightarrow H^0(X, S_{D_0}^{D_\nu})/V \longrightarrow H^1(X, \mathcal{O}_{D_0}) \longrightarrow H^1(X, \mathcal{O}_{D_\nu}) \longrightarrow 0 ,$$

so that $h^0(D_\nu) - h^0(D_0) = \dim V$ and $\deg D_\nu - \deg D_0 - \dim V = h^1(D_0) - h^1(D_\nu)$. Adding these two equations gives

$$\chi(D_\nu) - \chi(D_0) = \deg D_\nu - \deg D_0 ,$$

i.e. $\chi(D_\nu) - \deg D_\nu = \chi(D_0) - \deg D_0$ for $\nu = 1, 2$. The lemma follows.

We shall denote by g the integer $\dim_{\mathbb{C}} H^1(X, \mathcal{O})$; g is called the *genus* of the Riemann surface X.

RIEMANN–ROCH THEOREM: **Weak form**

If D is a divisor on the compact Riemann surface X, we have

$$\dim_{\mathbb{C}} H^0(X, \mathcal{O}_D) - \dim_{\mathbb{C}} H^1(X, \mathcal{O}_D) = \deg D + 1 - g ,$$

where g is the genus of X.

Proof. By Lemma 3, we have $\chi(D) - \deg D = \chi(0) - \deg 0$ (0 is the divisor 0). Now, $\chi(\mathcal{O}) = \dim H^0(X, \mathcal{O}) - \dim H^1(X, \mathcal{O}) = 1 - g$ (since the only holomorphic functions on all of X are the constants by the maximum principle).

Let D be a divisor on X. We shall denote by Ω_D the following sheaf on X: $\Omega_D(U) = \{$ meromorphic 1-form ω on $U \mid (\omega) \geq -D\}$, U open in X. Ω_D is isomorphic to the sheaf of holomorphic sections of $K_X \otimes L(D)$.

By the Serre duality theorem, we have $\dim_{\mathbb{C}} H^1(X, L(D)) = \dim_{\mathbb{C}} H^0(X, K_X \otimes L(D)^*) = \dim_{\mathbb{C}} H^0(X, K_X \otimes L(-D)) = \dim_{\mathbb{C}} H^0(X, \Omega_{-D})$. Thus, we may formulate the Riemann–Roch theorem as follows.

RIEMANN–ROCH THEOREM. *If D is a divisor on the compact Riemann surface X, we have*

$$\dim_{\mathbb{C}} H^0(X, \mathcal{O}_D) - \dim H^0(X, \Omega_{-D}) = \deg D + 1 - g .$$

Thus: the number of linearly independent meromorphic functions f on X with $(f) \geq -D$ is equal to $\deg D + 1 - g +$ *the number of linearly independent meromorphic 1-forms ω on X with $(\omega) \geq D$.*

We have seen (Theorem 3 in §7) that any holomorphic line bundle L on a compact Riemann surface X is isomorphic to $L(D)$ for some divisor D on X; moreover, two such divisors are linearly equivalent. We therefore define the degree $\deg(L)$ of the holomorphic line bundle L to be $\deg(D)$, where D is such that $L \simeq L(D)$, and write $h^i(L) = h^i(D)$.

By the Serre duality theorem, $h^0(K_X) = h^1(K_X^* \otimes K_X) = h^1(\mathcal{O}) = g$ (the genus) by definition of the genus.

Thus: there exist exactly g linearly independent holomorphic 1-forms on X.

By the Riemann–Roch theorem

$$h^0(K_X) - h^0(\mathcal{O}) = 1 - g + \deg K_X = g - 1\,;$$

thus: *The degree of the canonical line bundle K_X is $2g-2$. Equivalently, if $\omega \neq 0$ is any meromorphic 1-form, the degree of the divisor of ω is $2g-2$.*

One further remark: *if L is a holomorphic line bundle and $h^0(L) > 0$, then* $\deg(L) \geq 0$. *Further, if $h^0(L) > 0$ and* $\deg(L) = 0$, *then L is trivial.* In fact, if $s \in H^0(X, L)$, $s \not\equiv 0$, then $L \simeq L(D)$ where $D = \operatorname{div}(s) \geq 0$, so that $\deg L = \deg D \geq 0$. If $\deg L = 0$, then the divisor of s equals 0, i.e. s has no zeros and L is trivial.

This remark and the Serre duality theorem give us the following.

VANISHING THEOREM. Let D be a divisor on X and let $\deg(D) > 2g-2$. Then $H^1(X, \mathcal{O}_D) = 0$. If $\deg D > 0$, then $H^1(X, \Omega_D) = 0$.

Proof. $h^1(D) = h^0(K - D)$ (where K is any canonical divisor); since $\deg(K - D) = 2g - 2 - \deg(D) < 0$, $h^0(K - D) = 0$. Also, $\deg(\Omega_D) = 2g - 2 + \deg(D)$.

We shall denote by $\mathcal{M}$ the sheaf of meromorphic functions on X, i.e. the sheaf $U \mapsto \mathcal{M}(U) = \{\text{space of functions meromorphic on } U\}$; and by Ω_m the sheaf of meromorphic 1-forms on X.

Corollary. $H^1(X, \mathcal{M}) = 0$ and $H^1(X, \Omega_m) = 0$.

Proof. Let $\mathcal{U} = \{U_i\}_{i \in I}$ be a finite open covering of X, and let $\{V_i\}$ be open sets with $\bar{V}_i \subset U_i$ and $\cup V_i = X$. Let $\{f_{ij}\} \in Z^1(\mathcal{U}, \mathcal{M})$. We can clearly choose a divisor $D > 0$ with $\deg(D) > 2g-2$ such that $(f_{ij}) \geq -D$ on $V_i \cap V_j$ (because $\bar{V}_i \cap \bar{V}_j$ is compact in $U_i \cap U_j$. By the vanishing theorem, there are meromorphic functions f_i on V_i, $(f_i) \geq -D$ on V_i, with $f_i - f_j = f_{ij}$ on $V_i \cap V_j$. In particular, the image of $\{f_{ij}\}$ in $Z^1(\{V_i\}, \mathcal{M})$ lies in $B^1(\{V_i\}, \mathcal{M})$.

The proof for Ω_m is similar.

We can now give another interpretation of the duality pairing $H^0(X, \Omega_{-D}) \times H^1(X, \mathcal{O}_D) \to \mathbb{C}$.

Let $\{\omega_{ij}\} \in Z^1(\mathcal{U}, \Omega)$. Because of the corollary above, we can find meromorphic 1-forms ω_i on U_i such that $\omega_i - \omega_j = \omega_{ij}$ on $U_i \cap U_j$. Denote the cochain $\{\omega_i\} \in C^0(\mathcal{U}, \Omega_m)$ by

$\tilde{\omega}$. If $a \in X$, and i is such that $a \in U_i$, set $\mathrm{res}_a(\tilde{\omega}) = \mathrm{res}_a(\omega_i)$ [this is independent of i because $\omega_i - \omega_j$ is holomorphic on $U_i \cap U_j$].

We can also (Dolbeault isomorphism) find C^∞ forms of type (1,0) α_i on U_i such that

$$\omega_{ij} = \alpha_i - \alpha_j$$

and the 2-form φ on X given by $\varphi|U_i = \bar{\partial}\alpha_i = d\alpha_i$ represents the cohomology class $D(\{\omega_{ij}\})$ in $\mathcal{A}^{0,1}_{K_X}(X)$ under the Dolbeault isomorphism. We have

Lemma. $\int_X D(\{\omega_{ij}\}) = 2\pi i \sum_{a\in X} \mathrm{res}_a(\tilde{\omega})$.

Proof. Let $\beta = \omega_i - \alpha_i$ on U_i; β is a C^∞ form on $X - \{a_1, \ldots, a_n\}$, where $S = \{a_1, \ldots, a_r\}$ is the set of poles of $\tilde{\omega}$ ($S \cap U_i$ = set of poles of ω_i). Since $d\omega_i = 0$ on $U_i - S$, we have

$$\varphi = D(\{\omega_{ij}\}) = -d\beta \quad \text{on} \quad X - S\ .$$

Let $\Delta_{k,\varepsilon}$ be a small disc of radius ε around a_k $(k = 1, \ldots, r)$. Then

$$\int_X \varphi = \lim_{\varepsilon\to 0} \int\limits_{X - \bigcup_k \Delta_{k,\varepsilon}} d\beta = -\lim_{\varepsilon\to 0} \sum_k \int\limits_{\partial\Delta_{k,\varepsilon}} \beta\ .$$

Now, if $\varepsilon > 0$ is small, and i is such that $a_k \in U_i$, we have $\int\limits_{\partial\Delta_{k,\varepsilon}} \omega_i = 2\pi i\, \mathrm{res}_{a_k}(\omega_i) = 2\pi i\, \mathrm{res}_{a_k}(\tilde{\omega})$ and $\lim_{\varepsilon\to 0} \int\limits_{\partial\Delta_{k,\varepsilon}} \alpha_i = 0$ (since α_i) is smooth). Hence $\int_X \varphi = 2\pi i \sum_k \mathrm{res}_{a_k}(\tilde{\omega})$.

Thus, $\sum_k \mathrm{res}_{a_k}(\tilde{\omega})$ depends only on the cohomology class ξ of $\{\omega_{ij}\}$ in $H^1(X, \Omega)$ (and not on its representation as a coboundary of Ω_m); we denote this by $\mathrm{Res}(\xi)$.

Since $H^0(X, \mathcal{O}) = H^0(X, K_X^* \otimes K_X) = \mathbb{C}$, the pairing in the Serre duality theorem for $E = K_X$ is given by $(\lambda, \{\omega_{ij}\}) \mapsto \int \lambda \cdot D(\{\omega_{ij}\})$. Thus we have

Proposition 1. *The map* $\mathrm{Res} : H^1(X, \Omega) \to \mathbb{C}$ *is an isomorphism.*

Theorem. *(Residue version of Serre duality). Let D be a divisor on the compact Riemann surface X. We have a natural pairing $(\ ,\ )_D : H^0(X, \Omega_{-D}) \times H^1(X, \mathcal{O}_D) \to H^1(X, \Omega)$ defined by $(\omega, \{f_{ij}\}) \mapsto \{f_{ij}\omega\}$. [Here, ω is a meromorphic 1-form with $(\omega) \geq D$ and f_{ij} is meromorphic on $U_i \cap U_j$ with $(f_{ij}) \geq -D$, so that $f_{ij}\omega$ is a holomorphic 1-form on $U_i \cap U_j$].*

The duality pairing $\langle\ ,\ \rangle_{(D)}$ in the Serre duality theorem equals $2\pi i \mathrm{Res}(\ ,\ )_D$; in particular, the bilinear form

$$H^0(X, \Omega_{-D}) \times H^1(X, \mathcal{O}_D) \longrightarrow \mathbb{C}$$

$$(\omega, \xi) \longmapsto \mathrm{Res}\big((\omega, \xi)_D\big)$$

is non-degenerate.

As a consequence, we have the following analogue of Mittag–Leffler's theorem for compact Riemann surfaces.

Theorem. *Let $\{U_i\}_{i\in I}$ be an open covering of the compact Riemann surface X, and let f_i be meromorphic on U_i. Suppose that $f_i - f_j$ is holomorphic on $U_i \cap U_j$.*

There exists a meromorphic function f on X with $f - f_i$ holomorphic on $U_j \forall i$ if and only if, for any holomorphic 1-form ω on X, we have

$$\mathrm{res}(\tilde{\omega}_0) = 0 ,$$

where $\tilde{\omega}_0 \in C^0(\mathcal{U}, \Omega_m)$ is the 0-cochain $(f_i\omega)$.

[The condition is that if $\forall a \in X$ we choose $i(a)$ with $a \in U_i$, then $\sum_{a\in X} \mathrm{res}_a(f_{i(a)}\omega) = 0$.]

Proof. The existence of f is equivalent to saying that the cohomology class $\xi = \{f_i - f_j | U_i \cap U_j\}$ is $= 0$ in $H^1(X, \mathcal{O})$; by the above theorem, this is equivalent to the residue condition.

The corresponding theorem for meromorphic forms (rather than functions) is the following.

MITTAG–LEFFLER'S THEOREM for forms on a compact Riemann surface.

Let $\mathcal{U} = \{U_i\}_{i\in I}$ be an open covering of X, and let ω_i be a meromorphic 1-form on U_i such that $\omega_i - \omega_j$ is holomorphic on $U_i \cap U_j$. Let $\tilde{\omega} = \{\omega_i\}_{i\in I}$, and, for $a \in X$, set $\mathrm{res}_a(\tilde{\omega}) = \mathrm{res}_a(\omega_i)$ where i is such that $a \in U_i$. [This is independent of the i chosen since $\omega_i - \omega_j$ is holomorphic on $U_i \cap U_j$.]

Then, there exists a meromorphic 1-form ω on X with $\omega - \omega_i$ holomorphic on U_i $\forall i$ if and only if

$$\sum_a \mathrm{res}_a(\tilde{\omega}) = 0 .$$

Proof. The existence of ω is equivalent to saying that the cohomology class of $\{\omega_{ij}\}$ in $H^1(X, \Omega)$ is zero; since, by Proposition 1, the map $\mathrm{Res} : H^1(X, \Omega) \to \mathbb{C}$ is injective, this is equivalent to saying that $\sum_a \mathrm{res}_a(\tilde{\omega}) = 0$.

This can also be deduced from the duality theorem as follows (without knowledge of the precise duality pairing). Let $D \geq 0$, $D \neq 0$ be a non-zero effective divisor on X. Let s_D be the standard section of $L(D)$ with $(s_D) = D$. Consider the exact sequence of sheaves $0 \to \Omega \xrightarrow{s_D} \Omega_D \to \mathbb{C}_D \to 0$. The sheaf $\mathbb{C}_D$ is zero outside $\mathrm{supp}(D)$; if $a \in \mathrm{supp}(D)$, the stalk $\mathbb{C}_{D,a}$ can be identified with finite sums $\sum_{\nu=1}^{D(a)} \frac{c_\nu}{z^\nu}\, dz$ (where z is a local coordinate at a with $z(a) = 0$). The exact cohomology sequence is

$$H^0(X, \Omega_D) \longrightarrow H^0(X, \mathbb{C}_D) \longrightarrow H^1(X, \Omega) \longrightarrow H^1(X, \Omega_D)\,;$$

since $\dim H^1(X, \Omega) = 1$ and $H^1(X, \Omega_D) \simeq H^0(X, \mathcal{O}_{-D})^* = 0$, it follows that image $H^0(X, \Omega_D)$ has codimension 1 in $H^0(X, \mathbb{C}_D) \simeq \mathbb{C}^{\deg(D)}$. But, since if ω is a global 1 form, we have $\sum_a \operatorname{res}_a(\omega) = 0$, the image lies in the set

$$\{\omega_a\}_{a \in \operatorname{supp}(D)},\ \operatorname{ord}_a(\omega_a) \geq -D(a)$$

such that $\sum_{a \in \operatorname{supp}(D)} \operatorname{res}_a(\omega_a) = 0$. Since both spaces have codimension 1, they are equal.

The vanishing theorem $H^1(X, \mathcal{O}_D) = 0 = H^0(X, \Omega_{-D})$ if $\deg D > 2g - 2$ and the Riemann–Roch theorem show that $h^0(D) = \deg D + 1 - g$ is determined by the degree of D if it is large.

The integer $i(D) = h^1(D) = h^0(K - D)$ is called the *index of speciality* of D.

We give some further applications of these results.

Proposition 2. *If D is a divisor on X with $\deg D > 2g - 1$, then, $\forall P \in X$, $\exists s \in H^0(X, L(D))$ such that $s(P) \neq 0$. Equivalently, there is a divisor $D' \geq 0$, linearly equivalent to D, not containing P in its support.*

Proof. Let s_P be the standard section of $L(P)$ with $(s_P) = P$. The map

$$H^0(X, L(D - P)) \longrightarrow H^0(X, L(D))$$

given by $f \mapsto f \otimes s_P$ is not surjective (since, by the remark above, $h^0(D - P) = \deg(D - P) + (1 - g) < \deg D + (1 - g) = h^0(D)$ (since $\deg(D - P) > 2g - 2$). Its image consists precisely of sections of $L(D)$ vanishing at P.

Proposition 3. *Let L be a holomorphic line bundle on X with $degL > 2g$. Then*

(a) if $P, Q \in X$, $P \neq Q$, $\exists s \in H^0(X, L)$ such that $s(P) = 0$, $s(Q) \neq 0$

(b) if $P \in X$, $\exists s \in H^0(X, L)$ such that $\operatorname{ord}_P(s) + 1$.

Proof. We consider the bundle $L \otimes L(-P)$; since any line bundle is $\simeq L(D)$ for some divisor, Prop. 2 implies that $\exists s' \in H^0(X, L \otimes L(-P))$ with $s'(Q) \neq 0$. Let $s = s' \otimes s_P$, where s_P is the standard section of $L(P)$. Then, if $P \neq Q$, we have $s(Q) \neq 0$, $s(P) = 0$. If $P = Q$, we have $\operatorname{ord}_P(s_P) = 1$.

The imbedding theorem. Let L be a holomorphic line bundle on X with $\deg L > 2g$, and let $N = h^0(L) - 1 = \deg D - g$. We define a holomorphic map $\varphi_L : X \to \mathbb{P}^N$ as follows.

Let $s_0, \ldots, s_N$ be a basis of $H^0(X, L)$; if $a \in X$ choose a neighbourhood U of a and $\sigma \in H^0(U, L)$ with $\sigma(x) \neq 0\ \forall x \in U$. We set $\varphi_L(x) =$ point in projective space $\mathbb{P}^N$ with homogeneous coordinates $(\frac{s_0(x)}{\sigma(x)} : \ldots : \frac{s_N(x)}{\sigma(x)})$. Note that $\frac{s_j}{\sigma}$ is a holomorphic function on U; the point in $\mathbb{P}^N$ is independent of σ for if σ' is another such section and $\sigma' = h\sigma$ where h is a holomorphic function on U, $h(x) \neq 0 \forall x$, then $\frac{s_j}{\sigma} = h\frac{s_j}{\sigma'}$; φ_L is, to start with, defined only outside the common zeros of $s_0, \ldots, s_N$, but, by Prop. 2, these sections cannot have common zeros.

Note. If L is a holomorphic line bundle on X and $H^0(X, L) \neq 0$, it defines a holomorphic map $\varphi_L : X \to \mathbb{P}^N$ ($N = \dim H^0(X, L) - 1$) on *all* of X. If $s_0, \ldots, s_N$ is a basis as above and $A = \{x \in X \mid s_j(x) = 0\ \forall j\}$, the map is defined as above on $X - A$. If $a \in A$, and (U, z) is a small coordinate neighbourhood of a with $z(a) = 0$, $\varphi_L \mid U - \{a\}$ is the point in $\mathbb{P}^N$ with homogeneous coordinates $(f_0 : \ldots : f_N)$ where $f_0, \ldots, f_N$ are holomorphic functions on U and are not 0 outside a. We can write $f_j = z^k g_j$ where $k = \min_j \operatorname{ord}_a(f_j)$. φ_L on U is then given by the point in $\mathbb{P}^N$ with homogeneous coordinates $(g_0 : \ldots : g_N)$.

This construction only works because $\dim X = 1$; in higher dimension, points of indeterminacy of φ_L corresponding to the so-called *base points* of L, where all $s \in H^0(X, L)$ vanish, cannot be avoided in general.

We have: **The Imbedding Theorem.** *If* $\deg L > 2g$, *then* φ_L *is an imbedding of* X *in* $\mathbb{P}^N$.

Proof. (i) φ_L is injective. Let $P, Q \in X$, $P \neq Q$. Choose $s \in H^0(X, L)$ with $s(P) = 0$, $s(Q) \neq 0$; if $s = \sum_0^N c_\nu s_\nu$, then $\varphi_L(P)$ lies on the hyperplane $\sum c_\nu z_\nu = 0$, $\varphi_L(Q)$ lies outside this hyperplane.

(ii) The tangent map of φ_L is injective. Given $P \in X$, choose $s \in H^0(X, L)$ such that $\operatorname{ord}_P(s) = 1$, and let k, $0 \leq k \leq N$, be such that $s_k(P) \neq 0$, and let $c_0, \ldots, c_N \in \mathbb{C}$ be such that $s = \sum c_\nu s_\nu$. Then

$$\frac{s}{s_k} = c_k + \sum_{\nu \neq k} c_\nu \left(\frac{s_\nu}{s_k}\right),$$

and the functions $f_\nu = \frac{s_\nu}{s_k}$, $\nu \neq k$, give the inhomogeneous coordinates of $\varphi_L(x)$ for x near P: $\varphi_L(x) = \big(f_0(x), \ldots, 1, f_{k+1}(x), \ldots, f_N(x)\big)$. Since $\operatorname{ord}_P(\frac{s}{s_k}) = 1$, we have $df_\nu(P) \neq 0$ for at least one $\nu \neq k$.

A well known theorem of Chow implies that φ_L maps X onto the set of common zeros of finitely many homogeneous polynomials in the homogeneous coordinates of $\mathbb{P}^N$. Thus X is analytically isomorphic to a smooth algebraic curve in $\mathbb{P}^N$.

Moreover, an algebraic variety in $\mathbb{P}^N$ (irreducible and set of common zeros of finitely many homogeneous polynomials) if it is connected, of dimension 1 and a submanifold of $\mathbb{P}^N$, is obviously a compact Riemann surface. We shall therefore not distinguish between *compact Riemann surfaces* and (connected) *smooth projective curves.*

A holomorphic line bundle L on X is called *ample* if, for some integer $m > 0$, the mth-tensor power $L^{\otimes m}$ of L imbeds X in some projective space (i.e. if the corresponding map $\varphi_{L^{\otimes m}}$ is an imbedding of X in $\mathbb{P}^N$, $N + 1 = h^0(L^{\otimes m})$. It is called *very ample* if φ_L is already an imbedding.

We have seen that if $\deg(L) > 2g$, then L is very ample. Hence, if $\deg(L) > 0$, L is ample, since $\deg(L^{\otimes m}) = m \deg(L)$. Conversely, if L is ample, then $L^{\otimes m} \simeq L(D)$ for an *effective* divisor D (some $m > 0$) since $L^{\otimes m}$ must have at least one holomorphic section $\not\equiv 0$. Moreover, $D \neq 0$ (since then $L^{\otimes m}$ is trivial, and cannot imbed X). Thus $m \deg(L) = \deg(L^{\otimes m}) = \deg D > 0$.

Thus, we have

Proposition 4. *A holomorphic line bundle L on X is ample if and only if its degree is > 0.*

11. Further Properties of Compact Riemann Surfaces

Let X, Y be Riemann surfaces and $f : X \to Y$ a non-constant holomorphic map. If $a \in X$, $b = f(a)$ and w is a local coordinate at b with $w(b) = 0$, we set $\mathrm{ord}_a(f) = \mathrm{ord}_a(w \circ f)$. The integer $b(a, f) = \mathrm{ord}_a(f) - 1$ is called the ramification index of f at a; f is a local homeomorphism at a if and only if $b(a, f) = 0$.

Let now X, Y be compact Riemann surfaces, and let $f : X \to Y$ be a non-constant holomorphic map. We denote by g_X, g_Y the genera of X, Y respectively. Let $b = \sum_{a \in X} b(a, f)$; b is called the (total) ramification index of f. Let C be the set of critical points of f, i.e. $C = \{a \in X | b(a, f) > 0\}$ and $B = f(C)$ the set of critical values (sometimes called the branching locus).

Let $\omega \not\equiv 0$ be a meromorphic 1-form on Y and let $\omega_0 = f^*(\omega)$; we have $\deg(\omega_0) = 2g_X - 2$.

If $a \in X$, $b = f(a)$ and we choose local coordinates z at a and w at b ($z(a) = 0 = w(b)$) so that near a, the map f is given by $z \mapsto z^n = w$, then $n = \mathrm{ord}_a(f)$. If $\omega = h(w)dw$ near b, then $\omega_0 = f^*(\omega) = h(z^n) n z^{n-1} dz$ near a, so that

$$\mathrm{ord}_a(\omega_0) = n\, \mathrm{ord}_b(\omega) + n - 1\,, \quad \text{where} \quad n = \mathrm{ord}_a(f)\,.$$

If d is the number of sheets of f (= degree of f), this gives (summing first over $a \in f^{-1}(b)$ and then over b)

$$\begin{aligned}\deg(\omega_0) &= \sum_{b \in Y} \Big(\sum_{a \in f^{-1}(b)} \mathrm{ord}_a(f) \Big) \mathrm{ord}_b(\omega) + \sum_{b \in Y, a \in f^{-1}(b)} (\mathrm{ord}_a(f) - 1) \\ &= d \deg(\omega) + b\,.\end{aligned}$$

Since $\deg(\omega) = 2g_Y - 2$ and $\deg(\omega_0) = 2g_X - 2$, this gives:

The Riemann–Hurwitz Formula. With the above notation,

$$2g_X - 2 = d(2g_Y - 2) + b\,;$$

in particular, if there is a non-constant holomorphic map $X \to Y$, then $g_X \geq g_Y$; if $g_X = g_Y \geq 1$, we must have $b = 0$ and $d = 1$ unless $g_X = g_Y = 1$.

With the same notation as above, let $a_1, \ldots, a_r$ be the points of C, let $b_j = f(a_j)$. We denote by $\chi(X)$, $\chi(Y)$ the topological Euler characteristic of X, Y, so that, e.g.

$$\begin{aligned}&\chi(X) = \dim_{\mathbb{C}} H^0(X, \mathbb{C}) - \dim_{\mathbb{C}} H^1(X, \mathbb{C}) + \dim_{\mathbb{C}} H^2(X, \mathbb{C}) = 2 - b_1(X), \\ &b_1(X) = \dim_{\mathbb{C}} H^1(X, \mathbb{C}) = 1st \text{ Betti number of } X\,.\end{aligned}$$

We triangulate Y by simplices in which all points of $B = f(C)$ are vertices, and assume that the simplices are sufficiently small. We can then lift the triangulation by f to a triangulation of X. If we denote by $e_0(X)$ the number of vertices, by $e_1(X)$ the number of edges (= 1-simplices) and by $e_0(X)$ the number of faces (= 2-simplices) in the triangulation of X, with similar notation for the triangulation of Y, we have $e_2(X) = de_2(Y)$, $e_1(X) = de_1(Y)$, $e_0(X) = de_0(Y) - b$ [if $a_i \in C$, then each edge at $b_i = f(a_i)$ lifts to $b(a_i, f) + 1$ edges all ending in the same vertex a_i; the cardinality of $f^{-1}(B) = d$ (cardinality of B) $-b$].

Thus, we have

$$2 - b_1(X) = d\big(2 - b_1(Y)\big) - b\,.$$

If we take $Y = \mathbb{P}^1$, there exists a non-constant holomorphic map $f : X \to \mathbb{P}^1$ (= non-constant meromorphic function on X). Moreover, we have $g_{\mathbb{P}^1} = 0$, $b_1(\mathbb{P}^1) = 0$. If we denote by d the degree of f, we have

$$2g_X - 2 = -2d + b$$

and

$$2 - b_1(X) = 2d - b = -(2g_X - 2)\,.$$

Thus, we have

$$2g_X = b_1(X)\,;$$

in particular, the genus $g_X = \dim H^1(X, \mathcal{O}) = \dim H^0(X, \Omega)$ *is a topological invariant of* X.

We pass now to a discussion of Weierstrass points. Let X be a compact Riemann surface of genus $g = \dim H^1(X, \mathcal{O})$. We have seen (Th.3 in §7 and the remark following) that if $P \in X$, there is f meromorphic on X, holomorphic on $X - P$ (but not at P) with a pole of order $\leq g + 1$ at P.

It is natural to ask if this result can be improved and the order of the pole reduced; as we shall see, this is only possible for special choices of P (finite in number).

Given $P \in X$, let (U, z) be a coordinate neighbourhood at P with $z(P) = 0$. We call P a *Weierstrass point* if there is a meromorphic function f on X and constants $c_0, \ldots, c_{g-1}$, not all zero, such that

(i) $f|X - \{a\}$ is holomorphic

(ii) $f - \sum_{\nu=0}^{g-1} \frac{c_\nu}{z^{\nu+1}}$ is holomorphic at P.

According to the analogue of the Mittag–Leffler theorem given in §10, this is the case if and only if the following holds: There exist $c_0, \ldots, c_{g-1}$ not all zero such that

$$\operatorname{res}_P\Big(\sum_{\nu=0}^{g-1} \frac{c_\nu}{z^{\nu+1}}\omega\Big) = 0 \qquad \forall \omega \in H^0(X, \Omega)\,.$$

Let $\omega_1, \ldots, \omega_g$ be a basis of $H^0(X, \Omega)$, and let

$$\omega_k = f_k dz \quad \text{on} \quad U\ , f_k \in \mathcal{O}(U)\ .$$

Now, if $f_k = \sum_{\nu=0}^{\infty} f_{k,\nu} z^\nu$, we have $f_{k,\nu} = \frac{1}{\nu!}\left(\frac{d}{dz}\right)^\nu f_k|_{z=0} = \frac{1}{\nu!} f_k^{(\nu)}(0)$ and

$$\begin{aligned} \operatorname{res}_P\Big(\sum_{\nu=0}^{g-1} \frac{c_\nu}{z^{\nu+1}} \omega_k\Big) &= c_0 f_{k,0} + c_1 f_{k,1} + \cdots + c_{g-1} f_{k,g-1} \\ &= \sum_{\nu=0}^{g-1} \frac{c_\nu}{\nu!} f_k^{(\nu)}(0)\,. \end{aligned}$$

Thus we have: P is a Weierstrass point if and only if the system of g linear equations

$$\sum_{\nu=0}^{g-1} c_\nu f_k^{(\nu)}(0) = 0\ , \quad k = 1, \ldots, g$$

has a solution $(c_0, \ldots, c_{g-1}) \neq (0, \ldots, 0)$; this is the case if and only if $\det \left| \left(f_k^{(\nu)}(0)\right)_{\substack{1 \le k \le g \\ 0 \le \nu < g}} \right| = 0$.

We now make a few remarks on the Wronskian. Let U be a connected open set in $\mathbb{C}$ and $f_1, \ldots, f_n \in \mathcal{O}(U)$. Set $W(f_1, \ldots, f_n)(z) = \det\left(f_k^{(\nu)}(z)\right)\big|_{\substack{0 \le \nu < n \\ 1 \le k \le n}}$, $z \in U$. This is called the Wronskian of the functions $f_1, \ldots, f_n$. We have

Lemma. *$f_1, \ldots, f_n$ are linearly dependent over $\mathbb{C}$ if and only if $W(f_1, \ldots, f_n) \equiv 0$.*

Proof. If $c_1 f_1 + \cdots + c_n f_n \equiv 0$, $c_i \in \mathbb{C}$, $c_k \neq 0$, then the column $f_k^{(\nu)}$ $(0 \le \nu < n)$ is a linear combination of the columns $f_\ell^{(\nu)}$ $(0 \le \nu < n)$ with $k \neq \ell$ $\left[\text{viz} - \sum_{k \neq \ell} \frac{c_\ell}{c_k} f_\ell^{(\nu)} = f_k^{(\nu)}\right]$, so that the determinant is 0.

To prove the converse, we start with the following remark. Let $\varphi \in \mathcal{O}(U)$, $\varphi \not\equiv 0$ and let $g_k = \varphi f_k$, $k = 1, \ldots, n$. Then $g_k^{(\nu)} = \varphi f_k^{(\nu)} + \sum_{\mu < \nu} \lambda_\mu^\nu f_k^{(\mu)}$ $\left[\lambda_\mu^\nu = \binom{\nu}{\mu} \varphi^{(\nu-\mu)}\right]$ so that $\det\left(g_k^{(\nu)}\right) = \det\left(\varphi f_k^{(\nu)}\right) = \varphi^n \det\left(f_k^{(\nu)}\right)$ [the matrix $\left(g_k^{(\nu)}\right)$ is obtained from $\left(\varphi f_k^{(\nu)}\right)$ by adding multiples of rows $f_k^{(\mu)}$ $(1 \le k \le n)$ with $\mu < \nu$ to the νth row]. Thus, we have $W(g_1, \ldots, g_n) = \varphi^n W(f_1, \ldots, f_n)$.

We now prove that if $W(f_1, \ldots, f_n) \equiv 0$, then $f_1, \ldots, f_n$ are linearly dependent by induction on n. The result is trivial for $n = 1$ $\left(W(f_1) = f_1\right)$.

If $f_1 \not\equiv 0$ and $V = U - \{\text{set of zeros of } f_1\}$, it is enough to show that $1, \frac{f_2}{f_1}, \ldots, \frac{f_n}{f_1}$ are dependent (on V) over $\mathbb{C}$. Now, $W\left(1, \frac{f_2}{f_1}, \ldots, \frac{f_n}{f_1}\right) = f_1^{-n} W(f_1, \ldots, f_n) = 0$. If $g_k = \frac{f_k}{f_1}$

$(2 \le k \le n)$ then $W(1, g_2, \ldots, g_n) = W(\frac{dg_2}{dz}, \ldots, \frac{dg_n}{dz}) \equiv 0$. By induction, there are constants $c_2, \ldots, c_n$, not all zero, with $\sum_2^n c_k \frac{dg_k}{dz} \equiv 0$ on V, i.e. $\sum_2^n c_k g_k = \text{constant}$, so that $1, g_2, \ldots, g_n$ are linearly dependent over $\mathbb{C}$.

Returning to Weierstrass points, let $\omega_1, \ldots, \omega_g$ be a basis of $H^0(X, \Omega)$ as before, (U, z) a local coordinate. Set $W(\omega_1, \ldots, \omega_g) = W(f_1, \ldots, f_g)$ if $\omega_k = f_k dz$. Then, since the ω_k are $\mathbb{C}$-independent, $W(\omega_1, \ldots, \omega_g) \not\equiv 0$ on U, and we see that Weierstrass points are isolated, i.e. there are only finitely many Weierstrass points on X.

One further remark. If $w = w(z)$ is another coordinate system on U, so that $\omega_k = f_k dz = g_k(w) dw$, then $f_k = g_k(w(z)) \frac{dw}{dz}$, and we find that

$$f_k' = \left(\frac{dw}{dz}\right)^2 g_k'(w(z)) + \frac{d^2w}{dz^2} g_k(w(z))$$

$$\cdots\cdots\cdots\cdots\cdots\cdots\cdots\cdots$$

$$f_k^{(\nu)} = \left(\frac{dw}{dz}\right)^{\nu+1} g_k^{(\nu)}(w(z)) + \sum_{\mu<\nu} \lambda_\mu^\nu g_k^{(\mu)}(w(z))$$

with the λ_μ^ν independent of k. Thus $W(f_1, \ldots, f_n) = \left(\frac{dw}{dz}\right)^N W(g_1, \ldots, g_n)$ where $N = 1 + 2 + \cdots + g = \frac{1}{2}g(g+1)$.

Let (U_i, z_i) be an open covering of X by coordinate neighbourhoods. The transition functions $g_{ij} = \frac{dz_j}{dz_i}$ define the canonical bundle K_X of X, and the assignment $i \mapsto W_i = W(f_{1,i}, \ldots, f_{g,i})$ ($\omega_k = f_{k,i} dz_i$ on U_i) satisfies $W_i = g_{ij}^N W_j$. Thus, the W_i define a holomorphic section W of $K_X^{\otimes N}$, $N = \frac{1}{2}g(g+1)$. Thus we have the

Theorem. *There is a non-zero holomorphic section W of $K_X^{\otimes N}$, $N = \frac{1}{2}g(g+1)$, such that the zeros of W are exactly the Weierstrass points of X.*

Note that $\deg(\operatorname{div}(W)) = N \deg K_X = (g-1)g(g+1)$, so that there are at most $(g-1)g(g+1)$ Weierstrass points; if $g > 1$, W must have some zeros, and there must exist Weierstrass points. We also have the following:

WEIERSTRASS GAP THEOREM. Let X be a compact Riemann surface of genus $g > 0$, and let $P \in X$. Then, there exist exactly g integers $1 = n_1 < n_2 < \cdots < n_g \le 2g-1$ such that no meromorphic function on X, holomorphic on $X - P$, has a pole of order n_k, $k = 1, \ldots, g$, at P.

Thus, except for finitely many points P of X, there exists a meromorphic function on X, holomorphic on $X - P$, with a pole of order m at P if and only if $m \ge g+1$.

Proof. Let $D_k = k \cdot P$, $k = 0, 1, 2, \ldots$. There is f meromorphic on X, holomorphic on $X - P$, with a pole of order k at P if and only if $\exists f \in \mathbb{C}(X)$, $(f) \ge -D_k$, $(f) \not\ge -D_{k-1}$; if s_P is the standard section of $L(P)$ with $(s_P) = P$, this is equivalent to saying that

$$H^0(X, L(D_{k-1})) \xrightarrow{\otimes s_P} H^0(X, L(D_k))$$

is *not* surjective.

First, from the sequence $0 \longrightarrow \mathcal{O}_{D_{k-1}} \xrightarrow{s_P} \mathcal{O}_{D_k} \longrightarrow \mathcal{O}_{D_k}/\mathcal{O}_{D_{k-1}} \longrightarrow 0$, we obtain the exact sequence

$$0 \longrightarrow H^0\big(X, L(D_{k-1})\big) \longrightarrow H^0\big(X, L(D_k)\big) \longrightarrow H^0\big(X, \mathcal{O}_{D_k}/\mathcal{O}_{D_{k-1}}\big) .$$

Now, $\dim H^0(X, \mathcal{O}_{D_k}/\mathcal{O}_{D_{k-1}}) = 1$ (the quotient is the sheaf with $\mathbb{C}$ at P and 0 elsewhere). Hence, $\exists f$ with $(f) \geq -D_k$ but $(f) \not\geq -D_{k-1}$ if and only if $h^0(D_k) > h^0(D_{k-1})$, and in this case, $h^0(D_k) = 1 + h^0(D_{k-1})$.

Now, by the Riemann–Roch theorem,

$$h^0(D_k) - h^0(D_{k-1}) = 1 - h^1(D_{k-1}) + h^1(D_k) .$$

Hence, for any $m > 0$,

$$h^0(D_m) - h^0(D_0) = \sum_{k=1}^{m} \big(h^0(D_k) - h^0(D_{k-1})\big) = m - h^1(D_0) + h^1(D_m)$$

and $h^1(D_m) = 0$ for $m \geq 2g - 1$, while $h^0(D_0) = 1$, $h^1(D_0) = g$. Thus, for $m \geq 2g - 1$, we have

$$h^0(D_m) = m - g + 1 .$$

Moreover, $h^0(D_m) - 1 = \sum_1^m \big(h^0(D_k) - h^0(D_{k-1})\big)$ is the number of $k \leq m$ which occur as the order of pole at P of an $f \in \mathbb{C}(X)$ holomorphic on $X - P$. Thus, the number "gaps" is $m - \big(h^0(D_m) - 1\big) = g$.

Note. We could have applied this argument with an arbitrary sequence $P_1, P_2, \ldots$ of points and $D_0 = 0$, $D_k = \sum_1^k P_i$ for $k > 0$. We find that there is a meromorphic function f with $(f) \geq -D_k$ but $(f) \not\geq -D_{k-1}$ for all k except for g exceptional values $n_1, \ldots, n_g$ (between 1 and $2g - 1$) of k. This is sometimes called the Max *Noether gap theorem*.

12. Hyperelliptic Curves and the Canonical Map

Let X be a compact Riemann surface. We denote by $\mathbb{C}(X)$ the field of meromorphic functions on X.

We call X a *hyperelliptic Riemann surface*, or a *hyperelliptic curve* if it can be realised as a 2-sheeted (ramified) covering of $\mathbb{P}^1$, i.e. if there is $f \in \mathbb{C}(X)$ with either two simple poles or a single pole of order 2. We call such an f a function of degree 2. (**Note:** if there is $f \in \mathbb{C}(X)$ with one simple pole, then $f : X \to \mathbb{P}^1$ is an isomorphism since ∞ is not a critical value and $f : X \to \mathbb{P}^1$ is one-sheeted.) If X has genus $g > 0$, X is hyperelliptic if and only if there is an effective divisor D with $\deg(D) = 2$ and $h^0(D) \geq 2$. In fact, if X is hyperelliptic and f is of degree 2, we can take for D the divisor of poles of f [$h^0(D) \geq 2$ since $(1) \geq -D, (f) \geq -D$]. Conversely, if $h^0(D) \geq 2$, there is a non-constant f with $(f) \geq -D$.

If X has genus 1, it is hyperelliptic; in fact, if D is any divisor of degree 2, we have $\deg D > 2g - 2 = 0$, and $h^0(D) = 1 - g + \deg D = 2$. If X has genus 2, it is also always hyperelliptic; in fact, if P is a Weierstrass point, there is a function $f \in \mathbb{C}(X)$, holomorphic on $X - P$, with a pole of order 2 at P.

Let X be hyperelliptic and $f : X \to \mathbb{P}^1$ a function of degree 2. We can arrange that $f^{-1}(\infty)$ consists of two distinct points by composing f with an automorphism of P^1. If C is the set of critical points of f, and $B = f(C)$, we have $B \subset \mathbb{P}^1 - \{\infty\} = \mathbb{C}$. Moreover, $\mathrm{ord}_a(f) = 2$ if $a \in C$, so that $f^{-1}(b)$ consists of exactly one point if $b \in B$. The map $\tau : X - C \to X - C$ which takes x to the point $x' \neq x$ with $f(x) = f(x')$ extends to a holomorphic map $\tau : X \to X$ (by setting $\tau(a) = a$ for $a \in C$) and we have $\tau^2 =$ identity. It is called the hyperelliptic involution of X.

By the Riemann–Hurwitz formula, if X has genus g, we have since $\mathbb{P}^1$ has genus 0,

$$2g - 2 = -4 + \sum_{a \in C} (\mathrm{ord}_a(f) - 1) ,$$

so that (since $\mathrm{ord}_a(f) = 2$ for $a \in C$), the number of branch points of f is $2g + 2$ (and each of these is a Weierstrass point of X).

Let u be a meromorphic function on X such that $u(x) \neq u\big(\tau(x)\big)$ for some $x \in X - C$, u being holomorphic at both x and $\tau(x)$. Then, for all z near $f(x)$, $u(x_1) \neq u(x_2)$ if $(x_1, x_2) = f^{-1}(z)$. There exist rational functions $a_1, a_2 \in \mathbb{C}(z)$ so that (writing z for $f(x)$) we have

$$u^2(x) + 2a_1(z)u(x) + a_2(z) = 0$$

(proof of Theorem 4 in §7). Then $(u + a_1)^2 = a_1^2 - a_2 = p/q$, with $p, q \in \mathbb{C}[z]$; if we write $p \cdot q = P \cdot Q^2$ with $P, Q \in \mathbb{C}[z]$ and with P having no multiple roots, we have $w^2 = P(z)$, where $w = q(u + a_1)/Q$. Moreover, there is an open set of values of z such that $w(x_1) \neq w(x_2)$ if $(x_1, x_2) = f^{-1}(z)$.

Let Y be the Riemann surface (compact) of the algebraic function $w^2 - P(z)$. We obtain a map $\pi : X \to Y$ by setting $\pi(x) = \big(f(x), w(x)\big)$ outside a finite set on X (corresponding to the poles of f, w and the branch points of Y over $\mathbb{P}^1$, viz, the zeros of w); this extends to a holomorphic map $X \to Y$. Now, for an open set of $x \in X$, we have $w(x) \neq w\big(\tau(x)\big)$; since $w(x)^2 = w\big(\tau(x)\big)^2$, we have $w(x) = -w\big(\tau(x)\big)$, and, by the principle of analytic continuation, we have $w(x) = -w\big(\tau(x)\big)$ for all $x \in X$. It follows easily that π is an analytic isomorphism commuting with the projections $f : X \to \mathbb{P}^1$ and $(z, w) \mapsto z$ from Y to $\mathbb{P}^1$.

Moreover, since $f : X \to \mathbb{P}^1$ is not branched over ∞ and the number of branch points is $2g+2$, P is of the form $c(z - z_1) \dots (z - z_{2g+2})$ where $c \neq 0$ is constant, and $z_1, \dots, z_{2g+2}$ are distinct points in $\mathbb{C}$. We may assume that $c = 1$.

Consider now the 1-forms on Y defined by

$$\omega_\nu = z^{\nu-1}\frac{dz}{w}\,, \qquad \nu = 1, \dots, g\,.$$

Since $2wdw = P'(z)dz$ on Y, and $P'(z) \neq 0$ at the branch points $z_1, \dots, z_{2g+1}$, we have $\omega_\nu = 2z^{\nu-1}\frac{dw}{P'(z)}$ is holomorphic at points on Y over $\mathbb{P}^1 - \{\infty\}$. Near $z = \infty$, we have $w = \pm z^{g+1}\big(1 + O(\frac{1}{z})\big)$, so that $\omega_\nu = \pm z^{\nu-g-2}\,\big(1 + O(\frac{1}{z})\big)dz$ and this is holomorphic at $z = \infty$. Thus $\omega_1, \dots, \omega_g$ form a basis of $H^0(Y, \Omega)$. Also, we see that $\omega_0 \neq 0$ over $\mathbb{C} = \mathbb{P}^1 - \{\infty\}$ and $\omega_g \neq 0$ at the points over ∞.

Consider the holomorphic map $\varphi_{K_Y} = \varphi : Y \to \mathbb{P}^{g-1}$ given by the canonical bundle of Y. We see that $\varphi | Y - z^{-1}(\infty)$ is the map $(z, w) \mapsto (1, z, \dots, z^{g-1})$, and the image of Y is isomorphic to $\mathbb{P}^1$. Moreover $\varphi(z, -w) = \varphi(z, w)$. We have an isomorphism $\pi : X \to Y$ taking z to the function f. Moreover, the map φ_{K_X} determined by the canonical bundle of X is intrinsically defined, up to a linear transformation of $\mathbb{P}^{g-1}$; it is called *the canonical map of* X.

Thus, if $f : X \to \mathbb{P}^1$ is a function of degree 2, we see that it is isomorphic to the map $\varphi_{K_X} : X \to \varphi_{K_X}(X) \subset \mathbb{P}^{g-1}$ (and $\varphi_{K_X}(X) \simeq \mathbb{P}^1$). We conclude that f *is unique up to an automorphism of* $\mathbb{P}^1$, *i.e., that two functions of degree 2 differ only by a Möbius transformation* $f \mapsto \frac{af+b}{cf+d}$, $a, b, c, d \in \mathbb{C}$, $ad - bc \neq 0$. Now, if $f : X \to \mathbb{P}^1$ is of degree 2, then any branch point P of f is a Weierstrass point. This is obvious if $f(P) = \infty$; if $f(P) \neq \infty$, consider $\big(f - f(P)\big)^{-1}$.

We shall now show that conversely, any Weierstrass point on the hyperelliptic curve X is a branch point of the (essentially unique) map $f : X \to \mathbb{P}^1$ of degree 2, viz the canonical map. We identify X with the Riemann surface of $w^2 - P(z) = 0$, where

$P = (z - z_1) \cdots (z - z_{2g+2})$, the z_j being distinct. A basis of $H^0(X, \Omega)$ is given by $\omega_\nu = \frac{z^{\nu-1}}{w} dz$, $\nu = 1, \ldots, g$. If $w \neq 0$ (i.e. $P(z) \neq 0$), and $z \neq 0$ the Wronskian of $z^{\nu-1}/w$, $\nu = 1, \ldots, g$ equals $w^{-g} W(1, z, \ldots, z^{g-1}) = w^{-g} c_g$ (where $c_g = \prod_{\nu=1}^{g-1} (\nu!)$) [see proof of Lemma in §12, where we saw that $W(\varphi f_1, \ldots, \varphi f_n) = \varphi^n W(f_1, \ldots, f_n)$; also $W(1, z, \ldots, z^{g-1})$ is the determinant of a triangular matrix with $1, 1!, \ldots, (g-1)!$ on the diagonal]. If $z = \infty$, then $\omega_\nu = \pm z^{\nu-g-2}\{1 + O(\frac{1}{z})\} dz = \mp (\frac{1}{z})^{g-\nu}\{1 + O(\frac{1}{z})\} d(\frac{1}{z})$ and the Wronskian at ∞ is again the determinant of a triangular matrix with non-zero diagonal elements. Thus the points with $P(z) \neq 0$ and the points over $z = \infty$ are *not* Weierstrass points, proving our claim.

Thus, if X is hyperelliptic, the Weierstrass points are exactly the branch points of the canonical map $\varphi_{K_X} : X \to \varphi(K_X) \subset \mathbb{P}^{g-1}$. There are $2g + 2$ such points.

When $g > 2$, we have $2g+2 < (g-1) \cdot g \cdot (g+1)$, the bound on the number of Weierstrass points given before. It can be shown that non-hyperelliptic curves have more than $2g+2$ Weierstrass points.

For non-hyperelliptic curves, the canonical map is an imbedding.

Theorem. *Let X be a non-hyperelliptic compact Riemann surface of genus $g (\geq 3)$. Then, the canonical bundle K_X is very ample, i.e. global sections of K_X have no common zeros and $\varphi_{K_X} : X \to \mathbb{P}^{g-1}$ is an imbedding.*

Proof. 1) Given $P \in X$, $\exists \omega \in H^0(X, \Omega)$ with $\omega(P) \neq 0$. If this were false, the map $\Omega_{-P} \to \Omega$ given by tensoring with the standard section s_P of $L(P)$ would induce an isomorphism $H^0(X, \Omega_{-P}) \xrightarrow{\otimes s_P} H^0(X, \Omega)$. Now, $h^0(\Omega_{-P}) - h^1(\Omega_{-P}) = 1 - g + (2g - 3) = g - 2$ and $h^1(\Omega_{-P}) = h^0(\mathcal{O}_P) = 1$ (since, if there exists a non-constant f with $(f) \geq -P$, f has a single simple pole and $f : X \to \mathbb{P}^1$ is an isomorphism). Hence $h^0(\Omega_{-P}) = g - 1 < h^0(\Omega)$, so that $H^0(X, \Omega_{-P})$ cannot be isomorphic to $H^0(X, \Omega)$.

2) Given $P, Q \in X$, $P \neq Q$, $\exists \omega \in H^0(X, \Omega)$ with $\omega(P) = 0$, $\omega(Q) \neq 0$. If not, the map $H^0(X, \Omega_{-P-Q}) \xrightarrow{\otimes s_Q} H^0(X, \Omega_{-P})$ is an isomorphism. We have $h^0(\Omega_{-P-Q}) = 1 - g + (2g - 4) + h^1(\Omega_{-P-Q}) = g - 3 + h^0(\mathcal{O}_{P+Q})$. If $h^0(\mathcal{O}_{P+Q}) > 1$, there is a non-constant meromorphic function f with $(f) \geq -P - Q$, so that f is of degree 2 and X is hyperelliptic. Hence $h^0(\mathcal{O}_{P+Q}) = 1$, and $h^0(\Omega_{-P-Q}) = g - 2 < h^0(\Omega_{-P})$.

3) Given $P \in X$, $\exists \omega \in H^0(X, \Omega)$ with $\mathrm{ord}_P(\omega) = 1$. If not, we have $\omega(P) = 0 \Longrightarrow \mathrm{ord}_P(\omega) \geq 2$, i.e. $h^0(\Omega_{-P}) = h^0(\Omega_{-2P})$. As in 2) above, this implies the existence of a non-constant f with $(f) \geq -2P$ and X would be hyperelliptic.

The theorem follows from these three statements as in the imbedding theorem in §10.

The image of any compact Riemann surface X under φ_{K_X} is called the *canonical curve of X*. If X is hyperelliptic, the canonical curve is isomorphic to $\mathbb{P}^1$; otherwise, it is isomorphic to X.

13. Some Geometry of Curves in Projective Space

We begin with some general remarks. If M is a complex manifold of dimension n and $A \subset M$ is a submanifold of dimension $n-1$ (codimension 1), A defines a holomorphic line bundle as on a Riemann surface: if $\{U_i\}$ is an open covering of M, $f_i \in \mathcal{O}(U_i)$ is such that $U_i \cap A = \{x \in U_i | f_i(x) = 0,\ df_i \neq 0$ at any point of $U_i\}$, then $g_{ij} = f_i/f_j$ is holomorphic, nowhere zero on $U_i \cap U_j$ and form the transition functions for a line bundle $L(A)$. The family $\{f_i\}$ define the standard section s_A of $L(A)$ (whose divisor is A).

Consider now $M = \mathbb{P}^n$, with homogeneous coordinates $(z_0, \dots, z_n)$. A hyperplane H (linear subspace of codimension 1) is given by $\{\ell(z) = 0\}$, where ℓ is a non-zero linear form in $z_0, \dots, z_n$; we shall denote the corresponding line bundle also by H [or $\mathcal{O}_{\mathbb{P}^n}(1)$ or $\mathcal{O}(1)$]; two hyperplanes define isomorphic bundles. If $U_\nu = \{(z_0 : \dots : z_n) \mid z_\nu \neq 0\}$, $\nu = 0, \dots, n$, the functions $\left(\frac{z_0}{z_\nu}, \dots, \widehat{\frac{z_\nu}{z_\nu}}, \dots, \frac{z_n}{z_\nu}\right)$ form local coordinates on U_ν, in fact, an isomorphism onto $\mathbb{C}^n$ (the hat over a term means it is omitted). If H is defined by $\{\ell(z) = 0\}$, the functions $f_j = \frac{\ell(z)}{z_j}$ define $H \cap U_j$ $(j = 0, \dots, n)$ and the transition functions for H on $U_i \cap U_j$ are given by $g_{ij} = \frac{z_j}{z_i}$. The set of all hyperplanes forms the "dual" projective space $(\mathbb{P}^n)^*$, the coefficients of ℓ forming homogeneous coordinates for $(\mathbb{P}^n)^*$.

Let $X \subset \mathbb{P}^n$ be a connected complex submanifold of dimension 1 (= smooth imbedded projective algebraic curve). We set $\deg(X) = \deg(\mathcal{O}_{\mathbb{P}^n}(1)|X)$; it is called the degree of the curve X. If s_H is the standard section with divisor the hyperplane H, then $\mathcal{O}_{\mathbb{P}^n}(1)|X = \Sigma n_P P$ is the divisor of $s_H|X$ and $\deg(X) = \Sigma n_P$. If $X \cap H$ is transverse at every point, then $n_P = 1 \forall P \in X \cap H$, and $\deg(X)$ is the number of points in $X \cap H$.

A well-known theorem of Bertini implies that the "general" hyperplane meets X transversally; we shall prove it in the special case we need.

Proposition 1.* (Special case of Bertini's theorem). *The set of $H \in (\mathbb{P}^n)^*$ such that H meets X transversally is an open dense set in $(\mathbb{P}^n)^*$.*

Proof. Let $a \in X$ and U be a small open set given by a biholomorphic map $\varphi : \Delta \to U$, $\varphi = (\varphi_0, \dots, 1, \dots, \varphi_n)$, $\Delta = \{t \in \mathbb{C} | \, |t| < 1\}$ and $\varphi_0, \dots, \varphi_n$ are holomorphic functions

---

* If X is allowed to have singular points, the condition means that H avoids the singularities and is transverse elsewhere. The proposition holds also for such curves; if H_0 avoides the singularities of X, then all H near H_0 avoid an open set containing the singularities, and the proof applies.

with $\varphi_j'(t) \neq 0$ on Δ for some j and $\varphi_k \equiv 1$. Let $K \subset U$ be compact. For simplicity of notation, we suppose that $k = 0$. If $c_0 z_0 + \cdots + c_n z_n = 0$ is a hyperplane H, then $H \cap X$ is not transversal at some point of K if and only if $\sum_{\nu=1}^n c_\nu \varphi_\nu(t) = -c_0$ and $\sum_{\nu=1}^n c_\nu \varphi_\nu'(t) = 0$ have a common solution. Since $\varphi_\nu'(t) \not\equiv 0$ for some ν, we may assume that $\sum_{\nu=1}^n c_\nu \varphi_\nu'(t) \not\equiv 0$; the set $S = \{t \in U \mid \sum_{\nu=1}^n c_\nu \varphi_\nu'(t) = 0\}$ is discrete, and we can choose λ arbitrarily close to 0 so that $\sum_1^n c_\nu \varphi_\nu(t) \neq -c_0 - \lambda$ if $t \in S$; then $(c_0 + \lambda) z_0 + \cdots + c_n z_n = 0$ meets X transversally on K; thus the set $W_K = \{H | X \cap H$ is transversal at points of $K\}$ is dense. It is clearly open, and the set $\{H \in (\mathbb{P}^n)^*$ meeting X transversally$\}$ is a finite intersection of sets W_K.

In what follows, we shall assume that our curve $X \subset \mathbb{P}^n$ is *non-degenerate* in the sense that it is not contained in any hyperplane. One can always consider X imbedded in the smallest dimensional linear subspace $\mathbb{P}^k$ of $\mathbb{P}^n$ containing X.

Lemma 1. *If $X \subset \mathbb{P}^n$ is non-degenerate, then* $\deg(X) \geq n$.

Proof. Let H be a generic hyperplane and $X \cap H = \{x_1, \ldots, x_d\}$, $d = \deg(X)$. If $d < n$, choose points $y_1, \ldots, y_{n-d}$ on X. Now, any n points of $\mathbb{P}^n$ lie on a hyperplane. Let H' be a hyperplane containing $x_1, \ldots, x_d$, $y_1, \ldots, y_{n-d}$. Then, if $s_{H'}$ (standard section of $\mathcal{O}(1)$ with divisor H') is such that $s_{H'}|X \not\equiv 0$, we would have $\deg(s_{H'}|X) \geq n > d$, contradicting the definition of $\deg(X)$. Hence $s_{H'}|X \equiv 0$ and X is degenerate.

GEOMETRIC FORM OF THE RIEMANN–ROCH THEOREM. Let X be non-hyperelliptic and let $X \subset \mathbb{P}^{g-1}$ be the canonical imbedding. Let $D \geq 0$ be an effective divisor on X. Assume that $D \neq 0$. If ℓ is a linear form on $\mathbb{P}^{g-1}$, then $\ell|X$ is just a holomorphic 1-form on X and all such 1-forms are obtained in this way.

Given a hyperplane H in $\mathbb{P}^{g-1}$, we shall say that H contains D if the divisor of $\ell|X$, $(\ell|X) \geq D$, $\ell(z) = 0$ being an equation of H. We denote by $[D]$ the linear subspace $\mathbb{P}^{g-1}$ generated by D, which is, by definition, the intersection of all hyperplanes containing D. If D is of the form ΣP_i with the P_i distinct, a hyperplane containing D is just one containing the points P_i and $[D]$ is the subspace spanned by the P_i.

The geometric form of the Riemann–Roch theorem states simply that

$$h^0(D) = \deg(D) - \dim[D] .$$

In fact $h^0(\Omega_{-D})$ is the maximum number of linearly independent 1-forms ω on X with $(\omega) \geq D$, i.e. the maximum number of linearly independent linear forms ℓ on $\mathbb{P}^{g-1}$ with $(\ell|X) \geq D$, hence $g - 1 - h^0(\Omega_{-D})$ is the dimension of the intersection of all hyperplanes containing D, i.e. $g - 1 - h^0(\Omega_{-D}) = \dim[D]$, and the result follows from the Riemann–Roch theorem.

Before proceeding further, we shall need a very important theorem due to Castelnuovo.

Remark that any $k+1$ points ($k+1 \leq n$) in $\mathbb{P}^n$ lie on a k-dimensional linear subspace of $\mathbb{P}^n$. We say that $P_1, \ldots, P_{k+1} \in \mathbb{P}^n$ are (linearly) independent if they are not contained in a plane of dimension $< k$, i.e. if their linear span has the maximal possible dimension.

CASTELNUOVO'S GENERAL POSITION THEOREM. Let $X \subset \mathbb{P}^n$ be a non-degenerate algebraic curve of degree d. Then, the set of hyperplanes with the following property: if $X \cap H = \{x_1, \ldots, x_d\}$, then any set of n points $x_{i_1}, \ldots, x_{i_n}$, are linearly independent (i.e. do not lie in a plane of dimension $n-2$) is dense in $(\mathbb{P}^n)^*$.

In proving this result, we shall assume familiarity with basic algebraic geometry. We begin with the following.

Lemma 2. *Suppose that $n \geq 3$. Let $U \subset (\mathbb{P}^n)^*$ be the open set of hyperplanes meeting X transversally. (The condition that $H \in (\mathbb{P}^n)^*$ is not transverse to X is algebraic and U is the complement of a proper algebraic set in $(\mathbb{P}^n)^*$.) There is a proper algebraic subset A of U such that, if $H \in U - A$, then no three points of $X \cap H$ are colinear (lie on a line). [We suppose X is irreducible, but not necessarily smooth.]*

Proof. We shall prove that the (algebraic) family of lines in $\mathbb{P}^n$ which meet X in three (or more) points has dimension 1; these lines are called trisecants. Since the family of hyperplanes containing a given line has dimension $n-2$, this will show that hyperplanes containing some trisecant form a family of dimension $n-1$.

Suppose that trisecants form a family S of dimension ≥ 2. Since the family of all secants to X (family of lines joining two points on X) has dimension 2 and is irreducible (it is the closure of the image of $X \times X - \Delta_X$, $\Delta_X =$ diagonal, under the map sending P, Q to the line joining them), it would follow that every secant is a trisecant.

We first show that this implies that the tangent lines T_P and T_Q to X at any two smooth points P, Q must intersect.

Fix a point $P_0 \in X$ and consider the map $\pi_0 : X - P_0 \longrightarrow \mathbb{P}^{n-1}$ induced by projection of $\mathbb{P}^n$ to $\mathbb{P}^{n-1}$ from P_0. Let Y be the image curve and $y \in Y$ a smooth point such that π_0 is of maximal rank ($= 1$) at points of $\pi_0^{-1}(y)$. If $P, Q \in \pi_0^{-1}(y)$, then T_P, T_Q map onto the tangent line L of Y at y, so are contained in the plane generated by L and P_0 and so must intersect. Thus for fixed P_0 and an open set of points P, T_P, T_Q intersect if $\pi_0(P) = \pi_0(Q)$. If we vary P_0 in an open set, the corresponding points Q also fill out an open set. Thus there is an open set of pairs (P, Q) such that T_P, T_Q meet; this must then hold for all pairs (P, Q).

Choose P, Q with distinct tangent lines, and let B be the 2-plane spanned by T_P, T_Q. Now $B \cap X$ is finite (since X is non-degenerate), and let $a \in X$, $a \notin B$. Then, the tangent line T_a meets both T_P and T_Q, and since, $T_a \not\subset B$ and two lines meet in at most one point, T_a must contain the point $P_0 = T_P \cap T_Q$. Thus T_a contains P_0 for all but finitely many points a and hence $P_0 \in T_a$ $\forall a$. But this is absurd, since then projection from P_0 to $\mathbb{P}^{n-1}$ restricted to X, would be of rank 0 everywhere, its image would be a point, and X would actually reduce to a line.

This proves the lemma.

Proof of Castelnuovo's general position theorem. Let $U \subset (\mathbb{P}^n)^*$ be as in the lemma above and let $I \subset X \times U$ be the socalled incidence correspondence: $I = (P, H)$ with $P \in X \cap H$. Then I is irreducible of dimension n. [It is irreducible since projection on X has as fibre over $P \in X$ the irreducible family of hyperplanes through P; it is of dimension n because projection on U has finite fibres.]

Consider the subvariety $I_0 \subset I$ of pairs (P, H) such that there are points $P = P_1, \ldots, P_n \in X \cap H$ which are dependent. If $\dim I_0 < n$, then its projection on U would be proper, and the theorem, proved. Suppose $\dim I_0 = n$. Then, we have $I_0 = I$. If $P \in X$ is a general point and $\pi_P : X \to \mathbb{P}^{n-1}$ the projection from P, the general position theorem must be false for the image $X' \subset \mathbb{P}^{n-1}$ of X if it is false for X because if $P = P_1, P_2, \ldots, P_n \in H \cap X$ lie in a plane B of dimension $n-2$, then $\pi_P(P_2), \ldots, \pi_P(P_n)$ lie in $\pi_P(B)$ which has dimension $n-3$.

We can iterate this process as long as $n > 4$. Hence, it suffices to prove the theorem for $n = 3$. But in this case, the theorem is equivalent to Lemma 2.

Before giving some applications of the general position theorem we introduce some terminology.

Let X be a compact Riemann surface and L, a holomorphic line bundle on X. If V is a vector subspace of $H^0(X, L)$, $V \neq \{0\}$, we call the set of effective divisors $\{D | D = \operatorname{div}(s) \text{ for some } s \in V\}$, the *linear system* (or *series*) determined by V. If $V = H^0(X, L)$, we call it the *complete linear system* of L. If $L = L(D)$ for some divisor D, then, this complete linear system *consists of all effective divisors* $D' \geq 0$ *linearly equivalent to* D: $D' \sim D$, $D' \geq 0$. This is called the complete linear system of D and denoted by $|D|$. We shall write $\dim |D| = h^0(D) - 1$; it is called the dimension of the complete linear system, and $|D|$ is in $(1-1)$-correspondence with the projective space $(H^0(X, L(D)) - \{0\})/\mathbb{C}^* = \mathbb{P}(H^0(X, L(D)))$. If $h^0(D) > 0$ and $h^1(D) = h^0(\Omega_{-D}) > 0$, we call D a *special divisor*; this means that both D and $K_X - D$, where K_X is a canonical divisor on X, are linearly equivalent to effective divisors.

We begin with the following

Lemma 3. *Let D be a divisor with $h^0(D) > 0$, and let r be an integer ≥ 0. Then* $\dim |D| \geq r$ *if and only if, for any divisor $\Delta \geq 0$ of degree r, there is $D' \in |D|$ with $D' \geq \Delta$; in particular, if $P_1, \ldots, P_r \in X$, there is $D' \in |D|$ with $P_i \in$* $\operatorname{supp}(D')$ *for $i = 1, 2, \ldots, r$. If this condition holds for all P_i in a non-empty open set in X, the P_i being distinct, then* $\dim |D| \geq r$.

Proof. Suppose that $\dim H^0(X, L(D)) \geq r + 1$, and let $\Delta = \sum_1^k n_\nu P_\nu$. In terms of a local trivialization h_ν of $L(D)$ at P_ν, and local coordinates (U_ν, z_ν) at P_ν with

$z_\nu(P_\nu) = 0$, if $s \in H^0(X, L(D))$, then $(s) \geq \Delta$ if and only if $\left(\frac{d}{dz_\nu}\right)^\mu h_\nu(s)\,\big|_{s=P_\nu} = 0$ for $0 \leq \mu < n_\nu$, $\nu = 1, \ldots, k$. Thus, the condition is that s lie in the intersection of the kernels of the $n_1 + \cdots + n_k = r$ linear forms $s \mapsto \left(\frac{d}{dz_\nu}\right)^\mu h_\nu(s)\,\big|_{s=P_\nu}$ on $H^0(X, L(D))$, and this intersection has dimension $\geq \dim H^0(X, L(D)) - r \geq 1$.

The converse results from the following general fact.

Lemma 4. *Let X be a Riemann surface, L a holomorphic line bundle on X and V a vector subspace of $H^0(X, L)$ of dimension k. Then there are k points $P_1, \ldots, P_k \in X$ such that if $s \in V$ and $s(P_\nu) = 0$, $\nu = 1, \ldots, k$ then $s \equiv 0$. (In fact any k points in general position will do.)*

Proof. If $k > 0$, let $s_1 \in V$, $s_1 \not\equiv 0$, and let $P_1 \in X$ be so that $s_1(P_1) \neq 0$. Then $V_1 = \{s \in V | s(P_1) = 0\}$ is not all of V, so has dimension $k - 1$. If $k - 1 > 0$, choose $s_2 \in V_1$ and $P_2 \in X$ with $s_2(P_2) \neq 0$. Then $V_2 = \{s \in V_1 | s(P_2) = 0\} = \{s \in V | s(P_1) = 0, s(P_2) = 0\}$ has dimension $k - 2$. We have only to iterate this process.

A consequence is the following important

Proposition 2. *Let D_1, D_2 be divisors which are linearly equivalent to effective divisors (i.e. $h^0(D_i) > 0$, $i = 1, 2$). Then*

$$\dim |D_1| + \dim |D_2| \leq \dim |D_1 + D_2| \ .$$

Moreover, if equality holds, then any $D \in |D_1 + D_2|$ (i.e. $D \geq 0$, $D \sim D_1 + D_2$) can be written $D = D_1' + D_2'$ with $D_i' \in |D_i|$, $i = 1, 2$,.

Proof. If $r_i = \dim |D_i|$ and $P_1, \ldots, P_{r_1}$, $Q_1, \ldots, Q_{r_2}$ are any points in X, there is $D_i' \sim D_i$, $D_i' \geq 0$ with $P_i \in \operatorname{supp}(D_1')$, $Q_j \in \operatorname{supp}(D_2')$. Then $D_1' + D_2' \in |D_1 + D_2|$ and contains all $(r_1 + r_2)$ points P_i, Q_j in its support; hence the inequality.

The divisors $D_1' + D_2'$ with $D_i' \geq 0$, $D_i' \sim D_i$ form an $(r_1 + r_2)$-dimensional subvariety of the projective space $|D_1 + D_2| = \mathbb{P}(H^0(X, L(D_1 + D_2)))$. If equality holds, this subvariety must be the whole projective space.

We come now to an important theorem. We denote by K a canonical divisor on X.

CLIFFORD'S THEOREM. Let D be an effective special divisor on X (so that $h^0(K - D) > 0$). Let d be the degree of D. Then

$$\dim |D| \leq \frac{1}{2} d = \frac{1}{2} \deg D \ .$$

Moreover, if equality holds, then D must be 0, or $D \sim K$, or X must be hyperelliptic.

Proof. Since $K - D$ is linearly equivalent to an effective divisor, we have

$$\dim |D| + \dim |K - D| \leq \dim |K| \ ,$$

i.e.

$$h^0(D) + h^0(K - D) \le h^0(K) + 1 = g + 1 .$$

Moreover, by the Riemann–Roch theorem

$$h^0(D) - h^0(K - D) = d + 1 - g .$$

Adding, we have $2h^0(D) \le d + 2$, $h^0(D) \le \frac{1}{2}d + 1$. Moreover, if equality holds, then $\dim |D| + \dim |K - D| = \dim |K|$ so that any divisor $K' \ge 0$, $K \sim K'$, can be written $K' = D_1 + D_2$, $D_i \ge 0$ with $D_1 \sim D$, $D_2 \sim K - D$.

Assume that X is not hyperelliptic, and consider $X \subset \mathbb{P}^{g-1}$ imbedded by the canonical map. If H is any hyperplane transverse to X, then the points of $H \cap X$ give us a divisor $K' \sim K$, $K' \ge 0$, and we can write $K' = D_1 + D_2$ with $D_1 \sim D$, $D_2 \sim K - D$, $D_1, D_2 \ge 0$. Assume also that $D_1, D_2 \ne 0$.

If $[D_i]$ is the linear subspace of $\mathbb{P}^{g-1}$ generated by D_i, we have (by the geometric form of the Riemann–Roch theorem)

$$\begin{aligned} \dim[D_1] &= \deg D_1 - h^0(D_1) = d - h^0(D) \\ \dim[D_2] &= \deg D_2 - h^0(D_2) = 2g - 2 - d - h^0(K - D) . \end{aligned}$$

Since the assumption of equality $\dim |D| = \frac{1}{2}d$ implies that $h^0(D) + h^0(K - D) = g + 1$, this gives

$$\dim[D_1] + \dim[D_2] = 2g - 2 - (g + 1) = g - 3 .$$

Hence both D_1 and D_2 span linear subspaces of dimension $\le g - 3$. If $d \ge g - 1$, the points of D_1 are linearly dependent, if $d < g - 1$, the points of D_2 are linearly dependent. Since H is an arbitrary hyperplane transverse to X, this contradicts the general position theorem.

Thus, if X is not hyperelliptic, we must have D_1 or $D_2 = 0$, and the theorem is proved.

We now give another proof of Chifford's theorem not using Castelnuovo's general position theorem. We have only to prove the statement about equality.

Proposition 3. *Let $D \ge 0$ be an effective divisor of degree d. Assume that $0 \le d \le 2g - 2$. Then $\dim |D| \le \frac{1}{2}d$. If $D \ne 0$ and $D \not\sim K$, and if equality holds, then X is hyperelliptic.*

Proof. We have $h^0(D) - h^0(K - D) = 1 - g + d \le -\frac{1}{2}d + d$ (since $g - 1 \ge \frac{1}{2}d$) so that, if $h^0(K - D) = 0$, we have $\dim |D| \le \frac{1}{2}d - 1$. Thus, we may assume that D is special, in which case the inequality is a consequence of Prop. 2 (as in the first part of the above proof of Chifford's theorem), and in either case, $h^0(D) + h^0(K - D) \le g + 1$.

Assume that D is special and that $h^0(D) + h^0(K - D) = g + 1$ (i.e. that $h^0(D) = \frac{1}{2}d + 1$).

If $d = 2$, then $h^0(D) = 2$ and there is a non-constant meromorphic function f with $(f) \geq -D$ and f is of degree 2 so that X is hyperelliptic. We shall show that if $\deg D > 2$ and $K \not\sim D$, then there is a divisor $D_0 \geq 0$ with $\deg D_0 < d$ such that $h^0(D_0) + h^0(K - D_0) = g + 1$. Since $\deg D_0 < d \leq 2g - 2 = \deg K$, we have $K - D_0 \not\sim 0$ and we can continue till we obtain a divisor D' with $\deg D' = 2$, $h^0(D') = 2$, so that X is hyperelliptic.

Let $D' \geq 0$, $D' \sim K - D$. Then $D' \neq 0$. Choose points $P \in \operatorname{supp}(D')$, and $Q \notin \operatorname{supp}(D')$. Since $\dim |D| = \frac{1}{2}d > 1$, we can replace D by a linearly equivalent effective divisor whose support contains P and Q; we assume therefore that D has this property.

Let D_0 be the largest divisor $\leq D$ and $\leq D'$ (i.e. if $D = \sum_a D(a)a$, $D' = \sum_a D'(a)a$, then $D_0 = \sum_a \min(D(a), D'(a)) \cdot a$. Clearly, $D_0(P) > 0$, $D_0(Q) = 0$ so that $\deg D_0 < \deg D$ and $D_0 \neq 0$.

We have the following exact sequence of sheaves:

$$(*) \qquad 0 \longrightarrow \mathcal{O}_{D_0} \xrightarrow{\alpha} \mathcal{O}_D \oplus \mathcal{O}_{D'} \xrightarrow{\beta} \mathcal{O}_{D+D'-D_0} \longrightarrow 0\,,$$

where $\alpha(h) = (h, -h)$ and $\beta(f, g) = f + g$. The exactness is seen as follows. If $(h) \geq -D_0$, we have both $(h) \geq -D$ and $(h) \geq -D'$. If $(f) \geq -D$, $(g) \geq -D'$, we have $\operatorname{ord}_a(f + g) \geq -\max(D(a), D'(a)) = -\big(D(a) + D'(a) - \min(D(a), D'(a))\big)$, so that α, β are maps between the sheaves in question. If $\operatorname{ord}_a(f) \geq -\max(D(a), D'(a))$ (f a germ of meromorphic function at a), then either $(f) \geq -D(a)$ or $(f) \geq -D'(a)$; in the first case, $f = \beta(f, 0)$, in the second, $f = \beta(0, f)$. If $\beta(f, g) = 0$, then $f = -g$ and $(f) \geq -D$, $(f) = (g) \geq -D'$ so that $(f) \geq -D_0$ and $(f, g) = \alpha(f)$. Thus $(*)$ is exact.

It follows from the exact cohomology sequence that

$$h^0(D) + h^0(D') \leq h^0(D_0) + h^0(D + D' - D_0) = h^0(D_0) + h^0(K - D_0)\,.$$

Since $D' \sim K - D$, this gives

$$g + 1 = h^0(D) + h^0(K - D) \leq h^0(D_0) + h^0(K - D_0) \leq g + 1\,,$$

the last inequality following from the remark at the beginning of the proof.

This proves the existence of D_0, and hence, the proposition.

Corollary to Clifford's theorem. Let $X \subset \mathbb{P}^n$ be a compact Riemann surface of degree $d < 2n$ and suppose that X is non-degenerate. Then $g \leq d - n$ (g being the genus of X), and if equality holds, then the hyperplane sections form a complete linear system.

Proof. Let H be a hyperplane in $\mathbb{P}^n$ and $D = X \cap H$. Then $h^0(D) \geq n + 1$ (any linear form on $\mathbb{P}^n$ gives a section of $\mathcal{O}_D$, and no linear form vanishes on X unless it is zero since X is non-degenerate). Since $d < 2n$, we have

$$\dim |D| \geq n > \frac{1}{2}d\,,$$

so that D cannot be special, i.e. $h^0(K-D)=0$. Hence

$$n+1 \le h^0(D) = 1-g+d\,, \quad g \le d-n\,.$$

Equality implies that $h^0(D)=n+1$, i.e. restriction to X of linear forms on $\mathbb{P}^n$ give all sections of $\mathcal{O}_D$. This means, of course, that hyperplane sections form a complete linear system.

Corollary. A smooth non-degenerate curve of degree n in $\mathbb{P}^n$ is rational, i.e. $g=0$.

In fact, it can be shown that the only such curve is the closure of the image of $\mathbb{C}$ under the map $z \mapsto (1:z:z^2:\cdots:z^n)$ which we met as the canonical curve of a hyperelliptic Riemann surface.

Another application of the general position theorem was made by Castelnuovo himself to estimate the genus of a curve of degree $d >> n$ in $\mathbb{P}^n$.

Let $X \subset \mathbb{P}^n$ be a non-degenerate imbedding of a compact Riemann surface in $\mathbb{P}^n$; let d be the degree of X. Then, as we have seen $d \ge n$. Let $N = \left[\frac{d-1}{n-1}\right]$ (integral part). Let D be the divisor on X given by a general hyperplane section $X \cap H$.

Lemma 4. 1) *Let $1 \le k \le N$. Then $h^0(kD) - h^0\big((k-1)D\big) \ge 1 + k(n-1)$. Moreover, if equality holds for a certain value of k, then $H^0(X,\mathcal{O}_{kD})/H^0(X,\mathcal{O}_{(k-1)D})$ is generated by $H^0(X,D)$, i.e., the natural map $\operatorname{Sym}^k H^0(X,\mathcal{O}_D) \to H^0(X,\mathcal{O}_{kD})/H^0(X,\mathcal{O}_{(k-1)D})$ is surjective.*

2) *If $k > N$, we have $h^0(kD) - h^0\big((k-1)D\big) = d$, and $H^0(X,\mathcal{O}_D)$ generates $H^0(X,\mathcal{O}_{kD})/H^0(X,\mathcal{O}_{(k-1)D})$.*

Proof. We suppose that the hyperplane H is so chosen that it intersects X transversally and such that $D = X \cap H$ is in general position, i.e. that no n points of D lie on a plane of dimension $n-2$.

If $k \le N$, we have $k(n-1) \le d-1$, $1+k(n-1) \le d$. Choose a set E of $1+k(n-1)$ points of D.

If $P \in E$, write $E - \{P\} = E_1 \cup \cdots \cup E_k$ where each E_j has $n-1$ points. By Castelnuovo's general position theorem, the points of E_j $(j=1,\ldots,k)$ generate a plane B_j of dimension $n-2$ which does not contain P. Hence there is a hyperplane H_j with $P \notin H_j$, $E_j \subset H_j$, so that there is a linear form λ_j on $\mathbb{P}^n$ with $\lambda_j(P) \ne 0$, $\lambda_j(E_j)=0$. Let $\Lambda_P = \lambda_1 \ldots \lambda_k$; then Λ_P is a homogeneous polynomial of degree k with $\Lambda_P(P) \ne 0$, $\Lambda_P\big(E-\{P\}\big) = 0$. Let $s^{(P)} \in H^0(X,\mathcal{O}_{kD})$ be the section $\Lambda_P|X$. We claim that the images of the sections $s^{(P)}$, $P \in E$, in $H^0(X,\mathcal{O}_{kD})/H^0(X,\mathcal{O}_{(k-1)D})$ are linearly independent. In fact, If s_D is the standard section of $\mathcal{O}_D$ with divisor D, if $\{c_P\}_{P\in E}$ are complex numbers such that

$$\sum_{P\in E} c_P s^{(P)} \in s_D \cdot H^0\big(X,(k-1)D\big)\,,$$

then $\sum_{P\in E} c_P s^{(P)} = 0$ on D, hence on E; but the value of the sum $\sum c_P s^{(P)}$ on $Q \in E$ is $c_Q s^{(Q)}(Q)$ [since $s^{(P)}(Q) = 0$ if $Q \neq P$], so that, since $s^{(Q)}(Q) \neq 0$, we have $c_Q = 0$ $(\forall Q \in E)$. Hence $\dim H^0(X, Q_{kD})/H^0(X, \mathcal{O}_{(k-1)D}) \geq$ cardinality of $E = 1 + k(n-1)$. Since the sections $s^{(P)}$ are clearly $\in \mathrm{Sym}^k H^0(X, \mathcal{O})$ (since Λ_P is a product of linear forms), we have shown that the image of $\mathrm{Sym}^k H^0(X, \mathcal{O}_D)$ in $H^0(X, \mathcal{O}_{kD})/H^0(X, \mathcal{O}_{(k-1)D})$ has dimension $\geq 1+k(n-1)$. This proves both statements in part 1) of the lemma.

To prove part 2) we remark that if $k > \frac{d-1}{n-1}$, and $P \in \mathrm{supp}(D)$ we can write $\mathrm{supp}(D) - P = E_1 \cup \cdots \cup E_k$ where each E_j has at most $n-1$ points. As in the proof above, we can construct a homogeneous polynomial Λ_P of degree k, $\Lambda_P = \lambda_1 \cdots \lambda_k$ where λ_j is a linear form with $\lambda_j(P) \neq 0$, $\lambda_j(E_j) = 0$. Then, as in the proof above, the sections $s^{(P)} \in H^0(X, \mathcal{O}_{kD})$, $s^{(P)} = \Lambda_P | X$ are linearly independent, in $H^0(X, \mathcal{O}_{kD})/H^0(X, \mathcal{O}_{(k-1)D})$, and we find that $h^0(kD) - h^0\big((k-1)D\big) \geq d$.

On the other hand, the exact sequence

$$0 \longrightarrow \mathcal{O}_{(k-1)D} \xrightarrow{s_P} \mathcal{O}_{kD} \longrightarrow \mathbb{C}_D \longrightarrow 0$$

(where s_D is the standard section, and $\mathbb{C}_{D,x} = \mathcal{O}_{kD,x}/\mathcal{O}_{(k-1)D,x} = \mathbb{C}$ if $x \in D$, $= 0$ otherwise) implies that

$$h^0(kD) - h^0\big((k-1)D\big) \leq \dim H^0(X, \mathbb{C}_D) = d\,.$$

This proves 2) and also that for $k > N$, $H^0(X, \mathcal{O}_D)$ generates $\dfrac{H^0(X, \mathcal{O}_{kD})}{H^0\big(X, \mathcal{O}_{(k-1)D}\big)}$.

From this, we obtain

Castelnuovo's genus estimate. Let X be a (smooth) nondegenerate curve in $\mathbb{P}^n$. Let $d = \deg(X)$, and set $N = \left[\frac{d-1}{n-1}\right]$. Define ε $(0 \leq \varepsilon < n-1)$ by

$$d - 1 = N(n-1) + \varepsilon\,.$$

Then, the genus g of X is bounded as follows:

$$g \leq \frac{1}{2} N(N-1)(n-1) + N\varepsilon\,.$$

Further, if equality holds, then, for $k \geq 2$,

$$\mathrm{Sym}^k H^0(X, \mathcal{O}_D) \longrightarrow H^0(X, \mathcal{O}_{kD})$$

is surjective; i.e. $H^0(X, \mathcal{O}_D)$ generates $H^0(X, \mathcal{O}_{kD})$ for all $k \geq 2$.

Proof. Let r be a large positive integer. Then $h^1\big((r+N)D\big) = 0$, and by the Riemann–Roch theorem

$$h^0\big((r+N)D\big) = (r+N)d + 1 - g\,.$$

On the other hand

$$\begin{aligned} h^0\big((r+N)D\big) &= \sum_{k=1}^{N}\big(h^0(kD) - h^0\big((k-1)D\big)\big) + h^0(0.D) \\ &+ \sum_{k=N+1}^{N+r}\big(h^0(kD) - h^0\big((k-1)D\big) \\ &\geq \sum_{k=1}^{N}\big(1 + k(n-1)\big) + 1 + rd \quad \text{(by Lemma 4)} \\ &= 1 + rd + N + \frac{1}{2}N(N+1)(n-1)\ . \end{aligned}$$

It follows that

$$\begin{aligned} g &\leq (r+N)d - rd - N - \frac{1}{2}N(N+1)(n-1) \\ &= N(d-1) - \frac{1}{2}N(N+1)(n-1) \\ &= N^2(n-1) + \varepsilon N - \frac{1}{2}N(N+1)(n-1) = \frac{1}{2}N(N-1)(n-1) + \varepsilon N\ . \end{aligned}$$

Further, equality implies that $h^0(kD) - h^0\big((k-1)D\big) = 1 + k(n-1)$ for all $k \leq N$, and the fact that $H^0(X, \mathcal{O}_D)$ generates $H^0(X, \mathcal{O}_{kD})$ for all $k \geq 2$ follows by induction on k from the lemma. [Note that the function $1 \in H^0(X, \mathcal{O}_D)$, so that $\mathrm{Sym}^{k-1} H^0(X, \mathcal{O}_D) \subset \mathrm{Sym}^k\big(H^0(X, \mathcal{O}_D)\big)$.]

There are many beautiful geometric applications of this theorem of Castelnuovo. There is an excellent discussion of this circle of ideas in the book of Arbarello, Cornalba, Griffiths and Harris: *Geometry of Algebraic Curves*, I. (Springer-Verlag). We mention only one consequence, a famous theorem of Max Noether.

NOETHER'S THEOREM. Let X be a compact Riemann surface of genus $g \geq 3$. Suppose that X is not hyperelliptic. Then, if K_X is the canonical line bundle of X and $m \geq 2$, the natural map

$$\mathrm{Sym}^m H^0(X, K_X) \longrightarrow H^0(X, K_X^{\otimes m})$$

is surjective.

Proof. We consider $X \subset \mathbb{P}^{g-1}$ as the canonical curve. Then the hyperplane section D is a canonical divisor, hence $\deg D = \deg K = 2g-2$. The integer N above is $N = \left[\frac{2g-3}{g-2}\right] = 2$ if $g > 3$, $= 3$ if $g = 3$. If $g > 3$, $\varepsilon = 2g - 3 - 2(g-2) = 1$ and $\frac{1}{2}N(N-1)(n-1) + N\varepsilon = g - 2 + 2 = g$. If $g = 3$, $\varepsilon = 0$, $N = 3$ and

$\frac{1}{2}N(N-1)(n-1) + N\varepsilon = 3(g-2) = 3 = g$. Thus we have equality, and Noether's theorem follows from Castelnuovo's.

It should be added that if ($g \geq 3$ and) X is hyperelliptic, the above result definitely fails. This follows, e.g. from the fact that $K_X^{\otimes m}$ is very ample for large m, but the mapping φ_{K_X} induced by K_X is not injective.

14. Bilinear Relations

Before proceeding further, we recall some facts about compact oriented surfaces. We shall not prove them here; proofs can be found in, for example [6].

The basic theorem about the classification of compact orientable surfaces is the following:

A compact orientable C^∞ surface X without boundary is diffeomorphic to a sphere with a finite number of handles attached.

Attaching a handle is illustrated below.

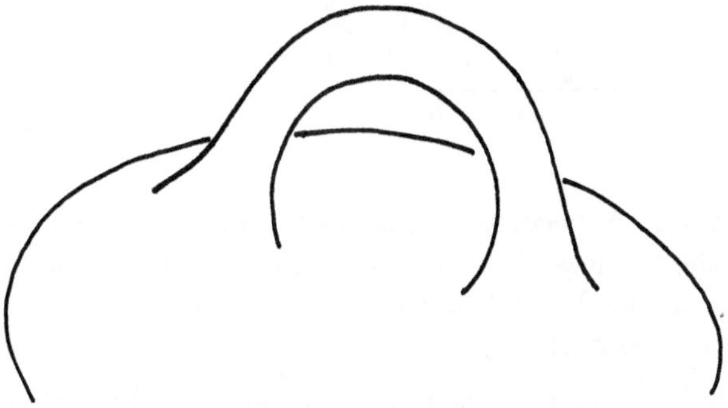

The number g of handles is half the first Betti number of X; thus, if X is a compact Riemann surface of genus g, it is diffeomorphic to a sphere with g handles, and two such surfaces are diffeomorphic.

A sphere with g handles can be described, up to diffeomorphism, as follows. Start with a convex polygon Δ with $4g$ sides $a_1, b_1, a_1', b_1', \dots, a_g, b_g a_g', b_g'$ in $\mathbb{C}$, oriented, as usual, "counter clockwise". If a_1, a_1' are the directed segments $\overline{pq}, \overline{p'q'}$, we identify a_1, a_1' by a linear map of $\overline{pq}$ onto $\overline{q'p'}$ (i.e. one taking p to q' and q to p'). Thus, a_1' is identified with a_1^{-1}. We make similar orientation reversing identifications of a_j with a_j' and of b_j with b_j' $(j = 1, \dots, g)$. This identification is indicated schematically below.

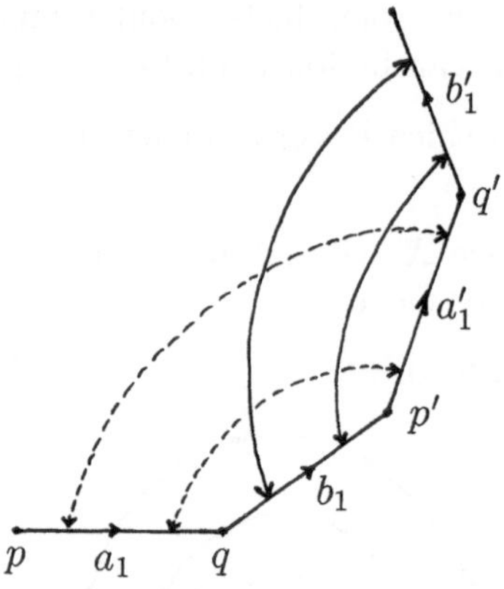

Under this identification, Δ becomes a compact surface X diffeomorphic to a sphere with g handles. All the vertices of Δ map onto the same point $x_0 \in X$, and a_j, b_j map onto closed curves at x_0 in X; we shall call these curves again a_j, b_j. The segments a_j', b_j' map onto a_j^{-1}, b_j^{-1} respectively.

These curves a_j, b_j in X form a basis of $H_1(X, \mathbb{Z})$ over $\mathbb{Z}$, and their intersection numbers are given by $a_i \cdot a_j = 0$, $b_i \cdot b_j = 0$, $a_i \cdot b_j = \delta_{ij} = -b_j \cdot a_i$ (δ_{ij} is the Kronecker δ; $\delta_{ij} = 1$ if $i = j$, 0 otherwise).

These curves are indicated schematically in the figure below.

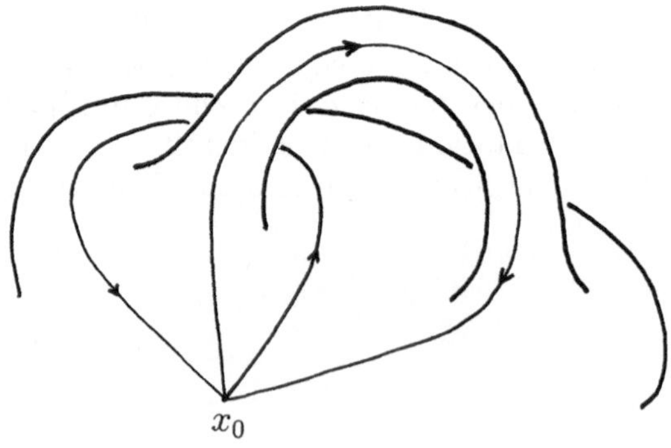

If we slit a sphere with g handles along curves a_j, b_j as shown (which have only x_0 as a point of intersection of any pair of them), we obtain a simply connected polygon Δ with $4g$ sides.

Let now X be a compact Riemann surface of genus g. We fix an identification (diffeomorphism preserving orientation) of X with a surface obtained from a $4g$ – gon Δ by the above identification process. This gives us (piecewise differentiable) curves a_i, b_j on X. If φ is a C^∞ 1-form defined in a neighbourhood of these curves, and is closed, we set

$$A_k(\varphi) = \int_{a_k} \varphi\,, \quad B_k(\varphi) = \int_{b_k} \varphi\,,$$

and call these the a- and the b-periods of φ.

Let α be a C^∞ closed 1-form on X, φ a C^∞ closed 1-form defined in a neighbourhood of $\bigcup a_i \cup \bigcup b_j$. We identify them with 1-forms on $\Delta(= \bar{\Delta})$ and on a neighbourhood of $\partial\Delta$ respectively. Fix $P_0 \in \mathring{\Delta}$ and, for $P \in \Delta$, set $u(P) = \int_{P_0}^{P} \alpha$ (Δ is simply connected). We then have

Lemma 1.

$$\int_{\partial\Delta} u\varphi = \sum_{k=1}^{g} (A_k(\alpha)B_k(\varphi) - B_k(\alpha)A_k(\varphi))\,.$$

Proof. Let $P \in a_k$ and let P' be the corresponding point of a'_k. Let γ be a curve joining P' to P as shown.

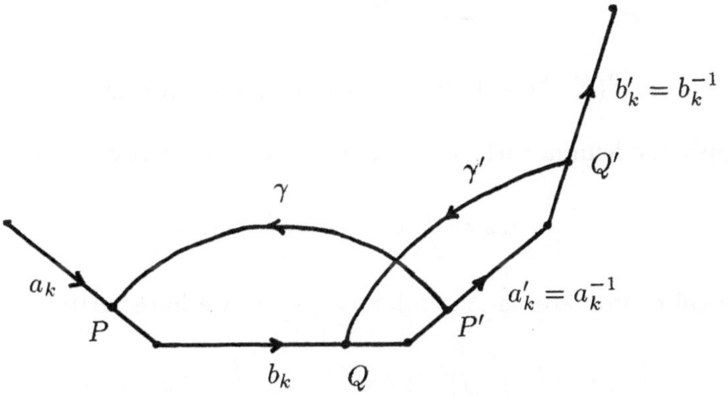

Then $u(P) - u(P') = \int_\gamma \alpha$; now, the image of γ in X is a closed curve homologous to b_k^{-1}, so that, since α is closed,

$$u(P) - u(P') = \int_{b_k^{-1}} \alpha = -B_k(\alpha)\,.$$

Similarly, if $Q \in b_k$ and Q' is the corresponding point on b'_k, we have

$$u(Q) - u(Q') = \int_{a_k} \alpha = A_k(\alpha)\,.$$

Now

$$\begin{aligned}\int_{\partial\Delta} u\varphi &= \sum_{k=1}^{g}\Big(\int_{a_k} + \int_{a'_k} + \int_{b_k} + \int_{b'_k}\Big)u\varphi \\ &= \sum_{k=1}^{g}\int_{a_k}(u(P)-u(P'))\varphi(P) + \sum_{k=1}^{g}\int_{b_k}(u(Q)-u(Q'))\varphi(Q)\end{aligned}$$

(with the notation above: $P \in a_k$, $Q \in b_k$ and P', Q' are the corresponding points on a'_k, b'_k respectively)

$$= \sum_{k=1}^{g}(-B_k(\alpha)\int_{a_k}\varphi + A_k(\alpha)\int_{b_k}\varphi)$$

which proves the lemma.

We deduce from this the following basic

Proposition 1. Let X be a compact Riemann surface of genus $g > 0$. We use the notation introduced above.

If ω is a holomorphic 1-form on X, $\omega \neq 0$, we have

$$\operatorname{Im}\sum_{k=1}^{g} A_k(\omega)\overline{B_k(\omega)} < 0\,.$$

In particular, if $\omega \in H^0(X,\Omega)$ and all its a-periods are 0, then $\omega = 0$.

Proof. We apply the lemma with $\alpha = \omega$, $\varphi = \bar\omega$. Now, by Stokes' theorem

$$\int_{\partial\Delta} u\bar\omega = \int_{\Delta} du \wedge \bar\omega = \int_X \omega\wedge\bar\omega\,.$$

If (U,z) is a local coordinate on X, and $z = x+iy$, we have setting $\omega = f\,dz$ on U, $f \in \mathcal{O}(U)$,

$$\int_U \omega\wedge\bar\omega = \int_U |f|^2\,dz\wedge d\bar z = -2i\int |f|^2\,dx\wedge dy\,,$$

so that $\frac{1}{2i}\int_X \omega\wedge\bar\omega < 0$ unless $\omega \equiv 0$. By the lemma,

$$\frac{1}{2i}\int_X \omega\wedge\bar\omega = \frac{1}{2i}\sum_{k=1}^{g}\big(A_k(\omega)\overline{B_k(\omega)} - B_k(\omega)\overline{A_k(\omega)}\big) = \operatorname{Im}\sum_{k=1}^{g} A_k(\omega)\overline{B_k(\omega)}\,.$$

Corollary. Let $\omega_1,\ldots,\omega_g$ be a basis of $H^0(X,\Omega)$. Let

$$A_{jk} = \int_{a_k}\omega_j\,.$$

Then, the matrix $(A_{jk})_{1\le j,k\le g}$ is invertible.

In fact if A_j is the g-vector $\left(\int_{a_1}\omega_j,\ldots,\int_{a_g}\omega_j\right)$, then, if $\Sigma c_j A_j = 0$, the a-periods of $\Sigma c_j\omega_j$ are zero, so that $\Sigma c_j\omega_j = 0$, $c_j = 0\ \forall j$.

In view of this corollary, we can choose a basis $\omega_1,\ldots,\omega_g$ of $H^0(X,\Omega)$ such that

$$\int_{a_k}\omega_j = \delta_{kj} \quad \text{(Kronecker } \delta) .$$

We shall call this a *normalized basis* of $H^0(X,\Omega)$ [relative to the choice a_i, b_j of basis of $H_1(X,\mathbb{Z})$].

Theorem. (Riemann's bilinear relations). *Let X be a compact Riemann surface of genus $g>0$. Let $\omega_1,\ldots,\omega_g$ be a normalized basis of $H^0(X,\Omega)$. Set*

$$B_{jk} = \int_{b_j}\omega_k .$$

Then, the complex matrix $B=(B_{jk})$ is symmetric, and its imaginary part is positive definite.

Proof. Let a_j, b_k, Δ have the same meaning as before, and let $u_j(P) = \int_{P_0}^{P}\omega_j$. We have

$$\int_{\partial\Delta} u_j\omega_k = \int_X \omega_j \wedge \omega_k \quad \text{(Stokes' theorem)} \ = 0 ;$$

on the other hand, (by Lemma 1),

$$\begin{aligned}\int_{\partial\Delta} u_j\omega_k &= \sum_{\nu=1}^{g}\left(A_\nu(\omega_j)B_\nu(\omega_k) - B_\nu(\omega_j)A_\nu(\omega_k)\right)\\ &= B_j(\omega_k) - B_k(\omega_j) \quad \text{since} \quad A_\nu(\omega_j) = \delta_{\nu j} .\end{aligned}$$

Thus, B is symmetric. Now, let $c_1,\ldots,c_g\in\mathbb{R}$, not all 0, and let $\omega = \sum_{k=1}^{g} c_k\omega_k$. By Prop. 1,

$$\operatorname{Im}\sum_{\nu=1}^{g} A_\nu(\Sigma c_k\omega_k)\overline{B_\nu(\Sigma c_k\omega_k)} < 0 .$$

Since $A_\nu(\Sigma c_k\omega_k) = c_\nu$, and $c_k\in\mathbb{R}$, this gives

$$\operatorname{Im}\sum_{\nu,k} c_\nu c_k \overline{B_{\nu k}} < 0 ,$$

i.e.

$$\operatorname{Im}\sum_{\nu,k} c_\nu c_k B_{\nu k} > 0 .$$

Given two distinct points P, Q, $P \neq Q$, on X, there is a meromorphic 1-form φ on X with simple poles at P and Q and $\mathrm{res}_P(\varphi) = +1$, $\mathrm{res}_Q(\varphi) = -1$ (by the Mittag–Leffler theorem for 1-forms given in §10). Because of the corollary to Prop. 1 above, we can add to φ a holomorphic 1-form φ' on X such that the a-periods of $\omega_{PQ} = \varphi + \varphi'$ are zero (we assume the a_i, b_j so chosen as not to contain P or Q); the form ω_{PQ} is then uniquely determined. It is called a *normalized abelian differential of the third kind.*

Given $P \in X$ an integer $n \geq 1$ and a coordinate system (U, z), at P with $z(P) = 0$, there is a unique meromorphic 1-form $\omega_P^{(n)}$ on X, holomorphic on $X - \{P\}$ and such that (i) $\omega_P^{(n)} - \frac{dz}{z^{n+1}}$ is holomorphic at D and (ii) the a-periods of $\omega_P^{(n)}$ are 0. This is called a *normalized abelian differential of the second kind.* (Abelian differentials of the first kind are simply holomorphic 1-forms. Any meromorphic 1-form on X is a linear combination of these three kinds of 1-form.)

RECIPROCITY THEOREM. Let ω_j, $j = 1, \dots, n$ be a normalized basis of $H^0(X, \Omega)$, and let $\omega_P^{(n)}$, ω_{PQ} be normalized abelian differentials of the 2nd and 3rd kind respectively. We have

1) $\int_{b_k} \omega_{PQ} = 2\pi i \int_Q^P \omega_k$ (the integral being taken along a curve joining P to Q in $X - \bigcup a_i - \bigcup b_j$)

2) If (U, z) is the coordinate system at P used to define $\omega_P^{(n)}$, and $\omega_k = f_k dz$, we have

$$\int_{b_k} \omega_P^{(n)} = 2\pi i \cdot \frac{1}{n!} f_k^{(n-1)}(P) .$$

Proof. We identify $X - \bigcup a_i - \bigcup b_j$ with a convex polygon Δ as before, and set $u_k(x) = \int_{P_0}^x \omega_k$ (P_0 a fixed point in Δ, the integral being along any path in Δ).
By Lemma 1, we have

$$\begin{aligned} \int_{\partial\Delta} u_k \omega_{PQ} &= \sum_\nu \{A_\nu(\omega_k) B_\nu(\omega_{PQ}) - B_\nu(\omega_k) A_\nu(\omega_{PQ})\} \\ &= B_k(\omega_{PQ}) \quad [\text{since } A_\nu(\omega_k) = \delta_{\nu k} \text{ and } A_\nu(\omega_{PQ}) = 0] \\ &= \int_{b_k} \omega_{PQ} . \end{aligned}$$

On the other hand, since Δ is simply connected and ω_{PQ} has residue +1 at P, -1 at Q, the residue theorem gives

$$\int_{\partial\Delta} u_k \omega_{PQ} = 2\pi i \big(u_k(P) - u_k(Q)\big) = 2\pi i \int_Q^P \omega_k .$$

This proves 1). The proof of 2) is similar: $\int_{b_k} \omega_P^{(n)} = \int_{\partial\Delta} u_k \omega_P^{(n)}$ (as above) $= 2\pi i \mathrm{res}_P\big(u_k \omega_P^{(n)}\big) = 2\pi i \mathrm{res}_P\big(u_k(z)\frac{dz}{z^{n+1}}\big) = 2\pi i \frac{1}{n!}\big(\frac{d}{dz}\big)^n u_k(z) = 2\pi i \frac{1}{n!} f_k^{(n-1)}(z)$ (since $\frac{du_k}{dz} = f_k$ near P).

The results in the reciprocity theorem are sometimes also referred to as bilinear relations for periods of differentials of the second and third kind. This aspect becomes clearer if we drop the normalization conditions we imposed above.

15. The Jacobian and Abel's Theorem

Let X be a compact Riemann surface of genus $g \geq 1$. We use the description of X in terms of a convex $4g$ – gon as in §14, and the corresponding basis a_i, b_j of $H_1(X,\mathbb{Z})$.

Let $\omega_1, \dots, \omega_g$ be a normalized basis of $H^0(X,\Omega)$: $\int_{a_j} \omega_k = \delta_{jk}$. Let Λ be the subgroup of $\mathbb{C}^g$ consisting of the vectors $\lambda_\gamma = \left(\int_\gamma \omega_1, \dots, \int_\gamma \omega_g\right)$ as γ runs over $H_1(X,\mathbb{Z})$. We have $\lambda_{a_k} = (0,\dots,1,\dots 0) = e_k$, the vector in $\mathbb{C}^g$ with 1 in the k-th place and 0 elsewhere, and $\lambda_{b_k} = \left(\int_{b_k} \omega_1, \dots, \int_{b_k} \omega_g\right)$ ($= B_k$ say) consists of the columns of the matrix $B = (B_{jk})$, $B_{jk} = \int_{b_j} \omega_k$. Since $\mathrm{Im}(B)$ is positive definite, the vectors $\{e_1,\dots,e_g,B_1,\dots,B_g\}$ are linearly independent over $\mathbb{R}$. Since $\{a_i, b_j\}$ generate $H_1(X,\mathbb{Z})$, we also have $\Lambda = \mathbb{Z}e_1 + \dots + \mathbb{Z}e_g + \mathbb{Z}B_1 + \dots + \mathbb{Z}B_g$. These remarks imply that Λ is a lattice in $\mathbb{C}^g$ with a compact quotient

$$J(X) = \mathbb{C}^g/\Lambda \,,$$

called the Jacobian of the Riemann surface X.

Intrinsically, $J(X)$ can be described as follows. Let V be the dual of $H^0(X,\Omega)$ (canonically isomorphic to $H^1(X,\mathcal{O})$ by the Serre duality theorem). We obtain a map of $H_1(X,\mathbb{Z})$ into V as follows if $\gamma \in H_1(X,\mathbb{Z})$, let its image in V be the linear form $\omega \mapsto \int_\gamma \omega$ on $H^0(X,\Omega)$. Then, the choices made above (of a_i, b_j and ω_k) identify $H^0(X,\Omega)^*$ with $\mathbb{C}^g$ and the image of $H_1(X,\mathbb{Z})$ with Λ. [The remarks made above also show that $H_1(X,\mathbb{Z})$ maps isomorphically onto a lattice in $H^0(X,\Omega)^*$.] We have

$$J(X) = H^0(X,\Omega)^* \;/\; \mathrm{im}\big(H_1(X,\mathbb{Z})\big) \,.$$

Fix a base point $P_0 \in X$. We define *the Abel–Jacobi map* $A : X \to J(X)$ as follows: choose a curve c from P_0 to P and set

$$A(P) = \Big(\int_{P_0}^{P} \omega_1, \dots, \int_{P_0}^{P} \omega_g\Big) \mod \Lambda$$

(where all integrals are along c). If c' is another curve from P_0 to P, there is an element $\gamma \in H_1(X,\mathbb{Z})$ with $\int_c \omega_k = \int_{c'} \omega_k + \int_\gamma \omega_k$ for all k, so that the map is well-defined. [In the intrinsic description, $A(P) =$ class of the linear form $\omega \mapsto \int_c \omega$ on $H^0(X,\Omega)$.]

Using the fact that $J(X)$ is an abelian group, we can define a map $X^N \to J(X)$ by $(P_1,\dots,P_N) \mapsto \sum_{j=1}^N A(P_j)$. Let $\mathrm{Div}(X)$ be the set of all divisors on X. We can also define a map $\mathrm{Div}(X) \to J(X)$ by

$$\sum_{i=1}^{r} n_i P_i \longmapsto \sum_{i=1}^{r} n_i A(P_i) \,.$$

We shall denote both these maps again by the same symbol A.

The theorem that is truly central in the study of the relationship between X and $J(X)$ is usually known as Abel's theorem. Abel's formulation of the theorem was rather different (and, in some ways, even more general). The theorem, as it is usually formulated today was first given by Riemann in his fundamental paper on abelian functions [2].

ABEL'S THEOREM. *Let D be a divisor on X of degree 0. Then D is linearly equivalent to 0 if and only if $A(D) = 0$ in $J(X)$.*

Thus, the theorem asserts the following. Let $P_1, \ldots, P_r; Q_1, \ldots, Q_r$ be points on X (with $Q_j \neq P_i \; \forall i, j$). The necessary and sufficient condition that there exist a meromorphic function with ΣP_i as its divisor of zeros and ΣQ_j as its divisor of poles (i.e. $(f) = P_1 + \cdots + P_r - Q_1 - \cdots - Q_r$) is that

$$\sum_{\nu=1}^{k} \int_{P_0}^{P_\nu} \vec{\omega} \equiv \sum_{\nu=1}^{k} \int_{P_0}^{Q_\nu} \vec{\omega} \bmod \Lambda \,, \quad \vec{\omega} = (\omega_1, \ldots, \omega_g) \,,$$

the integration being along some curve from P_0 to P_ν, and from P_0 to Q_ν respectively (the curves being, for each ν, the same for all ω_k).

Proof. Since D has degree 0, we can write $D = \sum_{k=1}^{r}(P_k - Q_k)$, the P_k, Q_k being points of X (and no Q_k being one of the P's).

Suppose there is a meromorphic function f with $(f) = D$. Then $\frac{df}{f} = \sum_{k=1}^{r} \omega_{P_k Q_k} + \sum_{\nu=1}^{g} c_\nu \omega_\nu$ with $c_\nu \in \mathbb{C}$. Further, $\int_\gamma \frac{df}{f} \in 2\pi i \mathbb{Z}$ $\forall$ closed curves γ on X (not containing any of P_k, Q_k).

Conversely, if $c_\nu \in \mathbb{C}$ are such that $\int_\gamma \varphi \in 2\pi i \mathbb{Z}$ $\forall$ closed curves γ, where $\varphi = \sum_{k=1}^{r} \omega_{P_k Q_k} + \sum_{\nu=1}^{g} c_\nu \omega_\nu$, then $(f) = D$ where $f(P) = \exp(\int_{P_0}^{P} \varphi)$ ($\exp(\int_{P_0}^{P} \varphi)$ is well-defined because of the condition on $\int_\gamma \varphi$).

We assume X identified with a convex polygon Δ as described earlier by slitting X along curves a_i, b_j which do not pass through P_k, Q_k. Let

$$\varphi = \sum_{k=1}^{r} \omega_{P_k Q_k} + \sum_{\nu=1}^{g} c_\nu \omega_\nu \,.$$

Then, $\int_\gamma \varphi \in 2\pi i \mathbb{Z}$ for all closed curves in $X - \bigcup\{P_k, Q_k\}$ if and only if $A_\nu(\varphi) = \int_{a_\nu} \varphi \in 2\pi i \mathbb{Z}$ and $B_\nu(\varphi) = \int_{b_\nu} \varphi \in 2\pi i \mathbb{Z}$ for all $\nu = 1, \ldots, g$; in fact, if C_k, C_k' denote small circles around P_k, Q_k respectively (with respect to coordinate neighbourhoods around these points), then γ is homologous to an integral linear combination of a_ν, b_ν, C_k, C_k', and $\int_{C_k} \varphi = +1$, $\int_{C_k'} \varphi = -1 \; \forall k$ since ω_{PQ} has residue $+1$ at P and -1 at Q.

Thus, there exists a meromorphic function f with $(f) = D$ if and only if:

(∗) There exist $c_\nu \in \mathbb{C}$ such that, if $\varphi = \sum_{k=1}^r \omega_{P_k Q_k} + \sum_{\nu=1}^g c_\nu \omega_\nu$, we have

$$A_\nu(\varphi),\ B_\nu(\varphi) \in 2\pi i \mathbb{Z},\ \nu = 1, \dots, g\,.$$

Now, $A_\nu(\varphi) = c_\nu$ because $\omega_{P_k Q_k}$ are normalized to have a-periods 0, and $A_\nu(\omega_\mu) = \delta_{\nu\mu}$. Moreover, by the reciprocity theorem

$$B_\nu(\varphi) = \sum_{k=1}^r 2\pi i \int_{Q_k}^{P_k} \omega_\nu + \sum_{\mu=1}^g c_\mu B_{\mu\nu}\,.$$

Thus, $A_\nu(\varphi)$, $B_\nu(\varphi)$ lie in $2\pi i\mathbb{Z}$ if and only if there exist integers $(n_1, \dots, n_g)$, $(m_1, \dots$ $\dots, m_g)$ such that

$$2\pi i\, n_\nu = c_\nu$$

and

$$\sum_{k=1}^r \int_{Q_k}^{P_k} \omega_\nu + \sum_{\mu=1}^g n_\mu B_{\mu\nu} = m_\nu$$

$(\nu = 1, \dots, g)$. This last condition can be written

$$\sum_{k=1}^r \int_{Q_k}^{P_k} \vec{\omega} = -\sum_{\mu=1}^g n_\mu B_\mu + \sum_{\mu=1}^g m_\mu e_\mu$$

e_μ is the vector with 1 in the μ-th place, 0 elsewhere, and B_μ is the vector $(B_{\mu 1}, \dots, B_{\mu g})$; since the e_μ, B_μ form a $\mathbb{Z}$-basis of Λ, we conclude that (∗) holds if and only if

$$\sum_{k=1}^r \int_{Q_k}^{P_k} \vec{\omega} \in \Lambda\,.$$

This proves the theorem.

We now study the relationship between $J(X)$ and X^g. Let S_n be the symmetric group on n letters [= group of permutations of $(1, \dots, n)$]. S_n acts on the cartesian product $X^n = \underbrace{X \times \cdots \times X}_{n\text{-times}}$; the quotient $S^n(X) = X^n/S_n$ is called the nth symmetric power of X. $S^n(X)$ is a complex manifold of dimension n. To introduce coordinates (especially at a point fixed by a non trivial element of S_n), we proceed as follows.

Consider the action of S_r on $\mathbb{C}^r$, and consider a neighbourhood of 0 in $\mathbb{C}^r$. Any germ of holomorphic function F at 0 in $\mathbb{C}^r$ invariant under S_r is a holomorphic function of the elementary symmetric functions in the coordinates $z_1, \dots, z_r$ of $\mathbb{C}^r$ (Newton's theorem); equivalently, it is a holomorphic function of $w_1 = z_1 + \cdots + z_r$, $w_2 = \frac{1}{2!}(z_1^2 + \cdots + z_r^2), \dots, w_r = \frac{1}{r!}(z_1^r + \cdots + z_r^r)$ and we can take $w_1, \dots, w_r$ as coordinates for $\mathbb{C}^r/S_r$.

If now $P_1, \ldots, P_n \in X$, and we number the P's so that $P_1 = \cdots = P_{r_1}$ ($= Q_1$ say), $P_{r_1+1} = \cdots = P_{r_1+r_2}$ ($= Q_2$ say), $\ldots, P_{r_1+\cdots+r_{p-1}+1} = \cdots = P_{r_1+\cdots+r_p}$ ($= Q_p$ say), $r_1 + \cdots + r_p = n$, and $Q_1, \ldots, Q_p$ are distinct, the group of elements of S_n which fix $(P_1, \ldots, P_n)$ is $S_{r_1} \times \cdots \times S_{r_p}$ (S_{r_k} acting by permutation on the kth block of r_k points), so that a neighbourhood of the image of $(P_1, \ldots, P_n)$ in $S^n(X)$ is isomorphic to a neighbourhood of $(0, \ldots, 0)$ in $\mathbb{C}^{r_1}/S_{r_1} \times \ldots \times \mathbb{C}^{r_p}/S_{r_p}$, and we can use the coordinates mentioned above for each factor.

Consider now the map $A : X^g \to J(X)$, $(P_1, \ldots, P_g) \mapsto \sum_{k=1}^g A(P_k)$. Clearly, A induces a map (which we shall again call A) $A : S^g(X) \to J(X)$. We shall identify $S^g(X)$ with the set of divisors $D \geq 0$ of degree g.

Theorem 1. *$A : S^g(X) \to J(X)$ is a birational map; there is an analytic set $Y \subset S^g(X)$ of dimension $< g$ such that $A : S^g(X) - Y \to J(X) - A(Y)$ is an analytic isomorphism.*

We start with

Lemma 1. *The set of distinct points $P_1, \ldots, P_g \in X$ such that the rank at $D = \Sigma P_i$ of the differential of $A : S^g(X) \to J(X)$ is maximal, $= g$, is open and dense in X^g.*

Proof. Let (U_j, z_j) be coordinates near P_j $\big(z_j(P_j) = 0\big)$. Using the local coordinates on $J(X)$ coming from $\mathbb{C}^g$ $\big(J(X) = \mathbb{C}^g/\Lambda\big)$, the map A can be written $A(z_1, \ldots, z_g) = \sum_j \int^{z_j} \vec{\omega}$ $\big(\vec{\omega} = (\omega_1, \ldots, \omega_g)\big)$. If $\vec{\omega} = \vec{f_j} dz_j$ on U_j $\big(\vec{f_j} = (f_{j_1}, \ldots, f_{j_g})\big)$ the Jacobian matrix of A at $D = \Sigma P_i$ is given by

$$\begin{pmatrix} \vec{f_1}(P_1) \\ \vdots \\ \vec{f_g}(P_g) \end{pmatrix}$$

The existence of $(P_1, \ldots, P_g)$ such that this matrix has rank g follows from §13, Lemma 4 (If $\dim H^0(X, L) = k$, $\exists k$ points x_j such that any $s \in H^0(X, L)$ vanishing at the x_j is identically 0) since $h^0(\Omega) = g$. The fact that this set is dense follows from the proof of §13, Lemma 4.

Lemma 2. *If $D = \sum_1^g P_i \in S^g(X)$, then $A^{-1}A(D)$ is the bijective holomorphic image of $\mathbb{P}^r$ with $r = \dim |D|$. In particular, $A^{-1}A(D)$ is connected $\forall D$.*

Proof. If $D_1, D_2 \in S^g(X)$ and $A(D_1) = A(D_2)$, then D_1 is linearly equivalent to D_2 by Abel's theorem (since $D_1 - D_2$ has degree 0). Let $\mathbb{P}^r = \mathbb{P}\big(H^0(X, \mathcal{O}_D)\big)$ $\big(D \in S^g(X)\big)$ be the projective space $\big(H^0(X, \mathcal{O}_D) - \{0\}/\mathbb{C}^*$. Because of the remark above, the map $H^0(X, \mathcal{O}_D) - \{0\} \to S^g(X)$, $s \mapsto \operatorname{div}(s)$ induces a bijection of $\mathbb{P}^r$ onto $A^{-1}A(D)$. The fact that this map is holomorphic can be seen as follows. Let $U \subset \mathbb{C}^N$ be open, and $f(x, t)$ a function holomorphic on $\Delta_\varepsilon \times U$ $[\Delta_\varepsilon = \{x \in \mathbb{C} \mid |x| < \varepsilon\}]$. Suppose that $f(x, t) \neq 0$ for $|x| = \rho (< \varepsilon)$, $t \in U$. Then, if $t_0 \in U$, the number of zeros $x_i(t)$ of $f(x, t)$

(counted with multiplicity) in $|x| < \rho$ is constant for t near t_0, say k, and, for $m \geq 0$, $m \in \mathbb{Z}$,

$$\sum_1^k (x_i(t))^m = \frac{1}{2\pi i} \int\limits_{|x|=\rho} x^m \frac{\frac{\partial f}{\partial x}(x,t)}{f(x,t)}\, dx \,,$$

so that this sum is holomorphic in t. We have only to apply this to the zeros of a general section $t_0 s_0 + \cdots + t_r s_r \in H^0(X, \mathcal{O}_D)$ (the s_j forming a basis) in view of the coordinates we are using on $S^g(X)$.

Note: The above map $\mathbb{P}^r \to A^{-1}A(D)$ is actually biholomorphic (see the corollary to Theorem 2 below).

Proof of Theorem 1. By Lemma 1, the set $Y = \{D \in S^g(X) \mid \operatorname{rank}(dA) \text{ at } D < g\}$ is an analytic set of dimension $< g$. By Lemma 2, $A^{-1}A(D) = \{D\}$ if $D \in S^g(X) - Y$. The result follows.

Remark that if $D \in S^g(X) - Y$, then $h^0(D) = 1$. In fact, Lemma 2 implies that if D is an isolated point in $A^{-1}A(D)$, then $\dim |D| = 0$. Since, by the Riemann–Roch theorem

$$h^0(D) - h^0(K - D) = 1 - g + \deg D = 1 \,,$$

we see that Y *consists exactly of the special divisors of degree* g.

Theorem 2. *For any* $D \in S^g(X)$*, the rank of the map* $A : S^g(X) \to J(X)$ *at* D *equals* $g - \dim |D|$.

Proof. Let $D = r_1 P_1 + r_2 P_2 + \cdots + r_n P_n$ with $r_j > 0$, $\Sigma r_j = g$ and $P_1, \ldots, P_n$ distinct. We take as coordinates at D on $S^g(X)$ the following functions [$x_1, \ldots, x_g$ being coordinates on X at $\underbrace{P_1, \ldots, P_1}_{r_1-\text{times}}, \ldots, \underbrace{P_n, \ldots, P_n}_{r_n-\text{times}}$ respectively]:

$$w_1^{(1)} = \sum_1^{r_1} x_i \,, w_2^{(1)} = \frac{1}{2!}\sum_1^{r_1} x_i^2, \ldots, w_{r_1}^{(1)} = \frac{1}{r_1!}\sum_1^{r_1} x_i^{r_1} \,,$$

$$w_1^{(2)} = \sum_{r_1 < i \leq r_2 + r_1} x_i \,, w_2^{(2)} = \frac{1}{2!} \sum_{r_1 < i \leq r_1 + r_2} x_i^2, \ldots, w_{r_2}^{(2)} = \frac{1}{r_2!} \sum_{r_1 < i \leq r_1 + r_2} x_i^{r_2}$$

and so on.

If $\omega_k = f_k dz$ near P_1, then, for $1 \leq i \leq r_1$, we have

$$\int_{P_0}^{x_i} \omega_k = \text{const} + x_i\, f_k(P_1) + \frac{x_i^2}{2!}\, f_k'(P_1) + \cdots + \frac{x_i^{r_1}}{r_1!}\, f_k^{(r_1-1)}(P_1) + \cdots \,.$$

Hence

$$\sum_{i=1}^{r_1} \int_{P_0}^{x_i} \omega_k = \text{const} + w_1^{(1)} f_k(P_1) + w_2^{(1)} f_k'(P_1) + \cdots + w_{r_1}^{(1)} f_k^{(r_1-1)}(P_1) + \mathcal{O}(w^2) \, .$$

Hence,

$$\sum_{i=1}^{g} \int_{P_0}^{x_i} \omega_k = \text{const} + \sum_{\nu=1}^{n} \sum_{1 \le j \le r_\nu} w_j^{(\nu)} f_k^{(j-1)}(P_\nu) + \mathcal{O}(w^2) \, .$$

Thus, the rank of the map A at D is that of the matrix with columns

$$\Phi_k = \begin{pmatrix} f_k(P_1) \\ \vdots \\ f_k^{(r_1-1)}(P_1) \\ f_k(P_2) \\ \vdots \\ f_k^{(r_2-1)}(P_2) \\ \vdots \end{pmatrix} \qquad k = 1, \ldots, g \, .$$

Now, a linear combination $\sum_{k=1}^{g} c_k \Phi_k$ of these column vectors is zero if and only if the holomorphic 1-form $\omega = \sum_{k=1}^{g} c_k \omega_k$ has the property that $\text{ord}_{P_\nu}(\omega) \ge r_\nu$ for $\nu = 1, \ldots, n$, i.e. if and only if $(\omega) \ge D$. Hence the number of linearly independent relations between the columns Φ_k is $h^0(\Omega_{-D})$; and since $h^0(D) - h^0(\Omega_{-D}) = 1 - g + g = 1$, we have $h^0(\Omega_{-D}) = h^0(D) - 1 = \dim |D|$. Hence the rank of the matrix $(\Phi_1, \ldots, \Phi_g)$, i.e. the rank of dA at D, is $g - \dim |D|$.

Corollary. For any $D \in S^g(X)$, $A^{-1}A(D)$ is a smooth submanifold of $S^g(X)$ and the map $\mathbb{P}\big(H^0(X, \mathcal{O}_D)\big) \to A^{-1}A(D)$ defined earlier is an analytic isomorphism.

In fact, Lemma 2 implies that $A^{-1}A(D)$ is an analytic set of dimension $\dim |D|$. That $A^{-1}A(D)$ is smooth follows from Theorem 2 and the implicit function theorem.

We have seen that the map $\mathbb{P}\big(H^0(X, \mathcal{O}_D)\big) \to A^{-1}A(D)$ is a holomorphic bijection between complex manifolds. It is a standard fact in complex analysis that such a map is biholomorphic.

We note two further consequences of these results. Let $\text{Div}(X)$ be the set of all divisors on X and let $P(X)$ be the subset of those divisors (of degree 0) which are linearly equivalent to 0. We set

$$\text{Pic}(X) = \text{Div}(X)/P(X) \, .$$

If $\text{Div}^0(X)$ is the set of all divisors of degree 0 on X, we set $\text{Pic}^0(X) = \text{Div}^0(X)/P(X)$. Then, we have

Theorem 3. *The Abel–Jacobi map $A : \mathrm{Div}(X) \to J(X)$ induces an isomorphism (of abelian groups)*

$$A : \mathrm{Pic}^0(X) \to J(X) .$$

Proof. That $A : \mathrm{Div}^0(X) \to J(X)$ is a homomorphism of abelian groups is clear. Abel's theorem asserts that the kernel of this map is exactly $P(X)$, so that we obtain the induced map $A : \mathrm{Pic}^0(X) \to J(X)$. We have only to prove that it is surjective. This follows from Theorem 1; if $D \in S^g(X)$ maps onto a given point $\zeta \in J(X)$, and P_0 is the base point in the definition of the Abel–Jacobi map $A : X \to J(X)$, then $D - gP_0$ has degree 0 and maps onto ζ.

Theorem 4. *If the genus $g > 0$, the Abel–Jacobi map $A : X \to J(X)$ is an imbedding.*

Proof. If $P, Q \in X$ and $A(P) = A(Q)$, then, by Abel's theorem, there is a meromorphic function f on X with $(f) = P - Q$, i.e. which has a single simple pole. As we have seen, this implies that X is isomorphic to $\mathbb{P}^1$, and the genus is 0. Thus, A is injective.

A is given by $\left(\int_{P_0}^x \omega_1, \ldots, \int_{P_0}^x \omega_g\right)$ and its tangent map at $P \in X$ is given by $\big(f_1(P), \ldots, f_g(P)\big)$, where $\omega_k = f_k dz$ in terms of a local coordinate. We have seen that the ω_k (hence the f_k) cannot all be zero at the same point, so that dA is also injective.

We end this section with a remark on Theorem 2. We have treated the case of the Abel–Jacobi map $S^g(X) \to J(X)$ because it is the most important. However, the theorem and its proof generalise as follows.

Theorem 5. *Let $1 \le k \le g$, and consider the map $A : S^k(X) \to J(X)$. If $D = P_1 + \cdots + P_k \in S^k(X)$, then the fibre $A^{-1}A(D)$ is a smooth submanifold, analytically isomorphic to $\mathbb{P}\big(H^0(X, \mathcal{O}_D)\big)$. The rank of the tangent map to A at D equals $k - \dim|D|$. Outside a proper analytic subset of $S^k(X)$, the map A is injective.*

If $k > g$, these statements, except the one about generic injectivity of A, remain true. The proof given when $k = g$ that the rank of dA at D is $g - \dim|D|$ extends to the general ease verbatim.

As for the injectivity statement, it is sufficient to show that the set $\{D \mid \text{rank of } dA \text{ at } D = k\}$ $(k \le g)$ is non-empty, i.e. the set $\{D \in S^k(X) \mid \dim|D| = 0\} \ne \emptyset$. But this follows from the fact that if $D' \ge 0$ is of degree g and $\dim|D'| = 0$, and we write $D' = D + D''$ with $\deg D = k$, and $D'' \ge 0$, then $\dim|D| = 0$.

16. The Riemann Theta Function

Let Λ be a lattice in $\mathbb{C}^g$, i.e. a subgroup of $\mathbb{C}^g$ which is discrete and of rank $2g$; the quotient $M = \mathbb{C}^g/\Lambda$ is a compact complex manifold, called a complex torus. Let L be a holomorphic line bundle on M, and $\pi : \mathbb{C}^g \to M$ the projection. A well-known theorem in complex analysis asserts that any holomorphic line (or even vector) bundle on $\mathbb{C}^g$ is holomorphically trivial. Let $h : \pi^*(L) \to \mathbb{C}^g \times \mathbb{C}$ be a trivialisation. If $\lambda \in \Lambda$ and $z \in \mathbb{C}^g$, then the isomorphisms $\pi^*(L)_z \to \mathbb{C}$ and $\pi^*(L)_{z+\lambda} \to \mathbb{C}$ differ by multiplication by a constant since $\pi^*(L)_z = \pi^*(L)_{z+\lambda} = L_{\pi(z)}$; if we denote this constant by $\varphi_\lambda(z)$, then for $\lambda \in \Lambda$, $z \mapsto \varphi_\lambda(z)$ is a holomorphic function without zeros, and we have, for $\lambda, \mu \in \Lambda$,

$$\varphi_\mu(z+\lambda)\varphi_\lambda(z) = \varphi_{\lambda+\mu}(z)\ , \quad z \in \mathbb{C}^g\ .$$

The family $\{\varphi_\lambda(z)\}$ is called a *factor of automorphy*. Conversely, any such family, i.e. any factor of automorphy, defines a holomorphic line bundle on M, obtained from $\mathbb{C}^g \times \mathbb{C}$ by identifying (z, u) and (w, v) if there is $\lambda \in \Lambda$ with $w = z + \lambda$ and $v = \varphi_\lambda(z)u$. A section of this line bundle can be interpreted as a holomorphic function f on $\mathbb{C}^g$ with $f(z+\lambda) = \varphi_\lambda(z)f(z)\ \forall\lambda \in \Lambda$. Such functions are called *multiplicative holomorphic functions.*

Let X be a compact Riemann surface of genus $g \geq 1$, and let $J(X) = \mathbb{C}^g/\Lambda$ be its Jacobian. We use the notation of §15, so that Λ has as a basis the vectors $e_k = (0, \ldots, 0, 1, 0, \ldots, 0)$ (1 in the k-th place) and $B_\nu = (B_{\nu 1}, \ldots, B_{\nu g})$, where $B_{\nu k} = \int_{b_\nu} \omega_k$.

There is a unique factor of automorphy $\{\varphi_\lambda\}$ with $\varphi_{e_k}(z) \equiv 1$, and $\varphi_{B_k}(z) = e^{-2\pi i z_k - \pi i B_{kk}}$, $k = 1, \ldots, g$.

Definition. Let $r \geq 1$ be an integer. A theta function of order r is a function θ holomorphic on $\mathbb{C}^g$ such that $\theta(z + e_k) = \theta(z)$, and $\theta(z + B_k) = e^{-2\pi i r(z_k + \frac{1}{2}B_{kk})}\theta(z)$, $k = 1, \ldots, g$.

Thus, a theta function of order r is a holomorphic section of $L^{\otimes r}$, where L is the line bundle on $J(X)$ defined by the factor of automorphy given by $\varphi_{e_k} = 1$, $\varphi_{B_k}(z) = \exp(-2\pi i z_k - \pi i B_{kk})$.

We now construct the *Riemann theta function*: it is given by

$$\vartheta(z) = \vartheta(z, B) = \sum_{n \in \mathbb{Z}^g} \exp\{\pi i \langle n, Bn\rangle + 2\pi i \langle n, z\rangle\}\ .$$

Here $B = (B_{\nu k})$ is the matrix $B_{\nu k} = \int_{b_\nu} \omega_k$; it is symmetric and has positive definite imaginary part. Moreover, if $z = (z_1, \ldots, z_g)$, $w = (w_1, \ldots, w_g)$, $\langle z, w\rangle = \Sigma z_i w_i$ is the standard bilinear form on $\mathbb{C}^g$.

Lemma 1. *The series defining $\vartheta(z)$ is uniformly convergent on compact subsets of $\mathbb{C}^g$, and ϑ is a theta function of order 1. Further, $\vartheta \not\equiv 0$, and $\vartheta(z) = \vartheta(-z)$.*

Proof. We have $\left|e^{\pi i\langle n,Bn\rangle}\right| = e^{-\pi\langle n,\mathrm{Im}(B)n\rangle}$. Now, since $\mathrm{Im}(B)$ is positive definite, there is $\delta > 0$ so that $\langle u, \mathrm{Im}(B)u\rangle \geq \delta\langle u,u\rangle = \delta\left|u\right|^2\ \forall u \in \mathbb{R}^n$. Thus

$$\left|e^{\pi i\langle n,Bn\rangle}\right| \leq e^{-\pi\delta|n|^2}\ , n \in \mathbb{Z}^g\ .$$

If K is compact in $\mathbb{C}^g$ there is a constant $C > 0$ such that

$$\left|e^{2\pi i\langle n,z\rangle}\right| \leq e^{C|n|}\ , z \in K\ .$$

The convergence follows.

Clearly $\vartheta(z + e_k) = \vartheta(z)$: ϑ is clearly periodic of period 1 in each variable, being a standard Fourier series. We have

$$\begin{aligned}\vartheta(z + B_k) &= \sum_{n\in\mathbb{Z}^g} e^{\pi i\langle n,Bn\rangle + 2\pi i\langle n,z\rangle + 2\pi i\langle n,B_k\rangle}(Be_k = B_k) \\ &= \sum_{n\in\mathbb{Z}^g} \exp\big(\pi i\langle(n+e_k),B(n+e_k)\rangle + 2\pi i\langle n+e_k,z\rangle \\ &\qquad\qquad - \pi i\langle e_k,Be_k\rangle - 2\pi i\langle e_k,z\rangle\big) \\ &= e^{-2\pi iz_k - \pi iB_{kk}}\vartheta(z)\ (\text{since } n+e_k \text{ runs over } \mathbb{Z}^g \text{ when } n \text{ does})\ .\end{aligned}$$

That $\vartheta \not\equiv 0$ follows from the fact that a Fourier series whose coefficients are not all zero cannot vanish identically. That $\vartheta(z) = \vartheta(-z)$ is obvious if we replace n by $-n$ in the series defining ϑ.

Lemma 2. *Any theta function of order 1 is a constant multiple of the Riemann theta function.*

Proof. Let $f(z)$ be a theta function of order 1. Since f is periodic, with period 1, in each variable, it has a Fourier expansion

$$f(z) = \sum_{n\in\mathbb{Z}^g} a_n e^{2\pi i\langle n,z\rangle}\ .$$

Now,

$$\begin{aligned}\Sigma a_n e^{2\pi i\langle n,z+B_k\rangle} &= f(z+B_k) \\ &= e^{-\pi iB_{kk} - 2\pi iz_k} f(z) = e^{-\pi iB_{kk}}\Sigma a_n e^{2\pi i\langle n-e_k,z\rangle} \\ &= \Sigma a_{n+e_k} e^{-\pi iB_{kk}} e^{2\pi i\langle n,z\rangle}\ .\end{aligned}$$

Hence $a_{n+e_k} = e^{\pi i B_{kk}+2\pi i\langle n,B_k\rangle} a_n$. It follows that if $a_n = 0$ for some n, then $a_{n+e_k} = 0$ $\forall k$, and hence that $a_n = 0$ for all n. In particular, $f \equiv 0$ if and only if $a_0 = 0$.

Applying this to $f - a_0\vartheta$, we conclude that $f \equiv a_0\vartheta$.

The Riemann theta function is a powerful tool in the study of the relationship between X and $J(X)$. The first use we shall make is to the proof of a famous imbedding theorem of Lefschetz. We begin with some preliminaries.

Let L be the line bundle on $J(X)$ defined by the factor of automorphy $\varphi_{e_k} \equiv 1$, $\varphi_{B_k} = e^{-2\pi i z_k - \pi i B_{kk}}$. Lemma 2 asserts that $H^0\big(J(X), L\big)$ has dimension 1, and ϑ defines a non-zero section of L. Let Θ be the divisor on $J(X)$ defined by this section: $\Theta = \operatorname{div}(\vartheta)$. Locally on $J(X)$, Θ is defined by the equation $\vartheta(z) = 0$; more precisely, if $a \in J(X)$ and $z_0 \in \mathbb{C}^g$ maps under the projection $\pi : \mathbb{C}^g \to J(X)$ onto a, if V is a small neighbourhood of z_0 and $\pi(V) = U$, $\Theta \cap U$ is defined by $\big(U, \vartheta \circ (\pi|V)^{-1}\big)$. Set theoretically, Θ is the image in $J(X)$ of $\{z \in \mathbb{C}^g \mid \vartheta(z) = 0\}$. It is called *the theta-divisor of* $J(X)$.

We need a slight generalisation of Lemma 2.

Lemma 3. Let r be an integer ≥ 1. The vector space V_r of theta functions of order r has dimension r^g; in particular it is finite dimensional.

Note: The finite dimensionality of $H^0(M, E)$, where M is a compact complex manifold and E, a holomorphic vector bundle on M, can be proved exactly as in the proof of §7, Theorem 1.

Proof of Lemma 3. Let $f \in V_r$; then, f is periodic of period 1 in each variable, and can be expanded in a Fourier series: $f(z) = \sum_{n\in\mathbb{Z}^g} a_n e^{2\pi i\langle n,z\rangle}$. we have

$$\sum_{n\in\mathbb{Z}^g} a_n e^{2\pi i\langle n,B_k\rangle} e^{2\pi i\langle n,z\rangle} = f(z+B_k) = e^{-2\pi i r z_k - \pi i r B_{kk}} f(z)$$

$$= \sum_{n\in\mathbb{Z}^g} a_n e^{-\pi i r B_{kk}} e^{2\pi i\langle n - re_k, z\rangle} = \sum_{n\in\mathbb{Z}^g} a_{n+re_k} e^{-\pi i r B_{kk}} e^{2\pi i\langle n,z\rangle}$$

so that $a_{n+re_k} = e^{\pi i r B_{kk}+2\pi i\langle n,B_k\rangle} a_n$. It follows at once that if $a_n = 0$ for those $n = (n_1, \ldots, n_g)$ with $0 \leq n_j < r$, then $f \equiv 0$. Hence $\dim V_r \leq r^g$.

Let $s = (s_1, \ldots, s_g) \in \mathbb{Z}^g$, $0 \leq s_j < r$. Define

$$\vartheta_{r,s}(z) = \sum_{n\in\mathbb{Z}^g} \exp\big\{\pi i\langle B(n + \frac{s}{r}), rn + s\rangle + 2\pi i\langle z, rn+s\rangle\big\}\,.$$

The series converges uniformly for z in any compact set in $\mathbb{C}^g$ as in Lemma 1, and one verifies, as in Lemma 1, that $\vartheta_{r,s} \in V_r \;\; \forall s$. Since the non-zero Fourier coefficients of $\vartheta_{r,s}$ are at the lattice points $\{s + rn | n \in \mathbb{Z}^g\}$ and these sets are pairwise disjoint for

$0 \le s_j < r$, it follows that $\{\vartheta_{r,s} \mid s = (s_1, \dots, s_g) \in \mathbb{Z}^g, 0 \le s_j < r\}$ are independent, hence form a basis of V_r.

Consider now a basis $\theta = (\theta_0, \dots, \theta_N)$, $N+1 = 3^g$ of the space of theta functions of order 3; as we shall see, the functions θ_j do not have common zeros; moreover, if $\lambda \in \Lambda$, $\theta(z+\lambda) = e^{w_\lambda(z)}\theta(z)$ where w_λ is a polynomial of degree ≤ 1 (as follows immediately from the definition of theta functions and the fact that e_k, B_k generate Λ over $\mathbb{Z}$). Hence, θ defines a holomorphic map, which we denote again by θ.

$$\theta : \mathbb{C}^g/\Lambda = J(X) \longrightarrow \mathbb{P}^N .$$

THE LEFSCHETZ IMBEDDING THEOREM. The map $\theta : J(X) \to \mathbb{P}^N$ defined by theta functions of order 3 is an imbedding.

Proof. We start by showing that V_2 *has no base points*, i.e. $\forall\ z_0 \in \mathbb{C}^g$, $\exists$ a theta function f of order 2 with $f(z_0) \neq 0$.

This follows from the following remark: if ϑ is Riemann's theta function and $a \in \mathbb{C}^g$, then $f(z) = \vartheta(z+a)\vartheta(z-a) \in V_2$. Also, if $a, b \in \mathbb{C}^g$, then $\vartheta(z+a)\vartheta(z+b)\vartheta(z-a-b)$ is a theta function of order 3. It follows that V_3 has no base points either (i.e. the basis functions $\theta_0, \dots, \theta_N$ have no common zeros).

We now show that $\theta : J(X) \to \mathbb{P}^N$ separates points. Suppose that $w_1, w_2 \in \mathbb{C}^g$ and that $\theta(w_1) = t\theta(w_2)$, $t \neq 0$. Then

$$\vartheta(w_1+a)\vartheta(w_1+b)\vartheta(w_1-a-b) = t\vartheta(w_2+a)\vartheta(w_2+b)\vartheta(w_2-a-b)\ \forall a, b \in \mathbb{C}^g .$$

We claim that this implies the following: the function $z \mapsto \frac{\vartheta(w_1+z)}{\vartheta(w_2+z)}$ is holomorphic and nowhere 0 on $\mathbb{C}^g$. In fact, given $z_0 \in \mathbb{C}^g$, we can choose $b \in \mathbb{C}^g$ so that $\vartheta(w_j+b) \neq 0$ and $\vartheta(w_j - z - b) \neq 0$ $j = 1, 2$ for all $z \in U$, where U is a small neighbourhood of z_0; we then have

$$\frac{\vartheta(w_1+z)}{\vartheta(w_2+z)} = \frac{t\vartheta(w_2+b)\vartheta(w_2-z-b)}{\vartheta(w_1+b)\vartheta(w_1-z-b)}, z \in U ,$$

and so is holomorphic and non-zero on U.

We now use the following lemma.

Lemma 4. *If* $w \in \mathbb{C}^g$ *is such that* $\frac{\vartheta(w+z)}{\vartheta(z)}$ *is holomorphic and nowhere* 0 *on* $\mathbb{C}^g$*, then* $w \in \Lambda$.

Equivalently, if $\zeta \in J(X)$ and the theta divisor Θ is left invariant by translation by $\zeta : \Theta = \Theta + \zeta$, then $\zeta = 0$ in $J(X)$.

Of course, this lemma and our remark above imply that $w_1 - w_2 \in \Lambda$, so that $\theta : J(X) \to \mathbb{P}^N$ is injective.

Proof of Lemma 4. There exists a holomorphic function g on $\mathbb{C}^g$ such that

$$\frac{\vartheta(w+z)}{\vartheta(z)} = e^{g(z)}\ , z \in \mathbb{C}^g\ .$$

Since ϑ is periodic with period 1 in each variable, there exists, for $1 \le k \le g$, an integer n_k such that

$$g(z+e_k) - g(z) = 2\pi i n_k\ , k = 1, \ldots, g\ .$$

Further,

$$\begin{aligned} \exp\bigl(g(z+B_k)\bigr) &= \frac{e^{-2\pi i(z_k+w_k)-\pi i B_{kk}}\vartheta(w+z)}{e^{-2\pi i z_k - \pi i B_{kk}}\vartheta(w)} \\ &= e^{-2\pi i w_k}\exp\bigl(g(z)\bigr)\ . \end{aligned}$$

Hence, there exist integers m_k such that

$$g(z+B_k) - g(z) = -2\pi i w_k + 2\pi i m_k\ .$$

For any ν $1 \le \nu \le g$, it follows that

$$\frac{\partial g}{\partial z_\nu}(z+\lambda) = \frac{\partial g}{\partial z_\nu}(z) \quad \text{if} \quad \lambda = e_k \quad \text{or} \quad \lambda = B_k\ , k = 1, \ldots g\ ;$$

hence

$$\frac{\partial g}{\partial z_\nu}(z+\lambda) = \frac{\partial g}{\partial z_\nu}(z) \quad \forall \lambda \in \Lambda\ ,$$

so that $\frac{\partial g}{\partial z_\nu}$ defines a holomorphic function on the compact connected manifold $J(X)$, and so is constant. Hence, there exist constants $c_0, c_1, \ldots, c_g$ so that

$$g(z) = c_0 + c_1 z_1 + \cdots + c_g z_g\ .$$

Hence $g(z+e_k) - g(z) = c_k = 2\pi i n_k$, and

$$\begin{aligned} 2\pi i w_k &= -\bigl(g(z+B_k) - g(z)\bigr) + 2\pi i m_k \\ &= -\sum_\nu c_\nu B_{\nu k} + 2\pi i m_k = -2\pi i \sum_\nu n_\nu B_{\nu k} + 2\pi i m_k\ . \end{aligned}$$

Thus $w = -\sum_\nu n_\nu B_\nu + \sum_k m_k e_k \in \Lambda$ as desired.

We next show that the tangent map $d\theta$ of θ is also injective, i.e. that $\theta : J(X) \to \mathbb{P}^N$ is an immersion.

If $a \in \mathbb{C}^g$, the injectivity of $d\theta$ at $\pi(a)$ [$\pi : \mathbb{C}^g \to J(X)$ being the projection] is equivalent to the following statement: the rank of the matrix

$$\begin{pmatrix} \theta_0(a) & \ldots & \theta_N(a) \\ \frac{\partial \theta_0}{\partial z_1}(a) & \ldots & \frac{\partial \theta_N}{\partial z_1}(a) \\ \ldots & \ldots & \ldots \\ \frac{\partial \theta_0(a)}{\partial z_g} & \ldots & \frac{\partial \theta_N}{\partial z_g}(a) \end{pmatrix}$$

equals $g+1$.

Suppose that the rank were $< g+1$. Then, there exist $c_0, \ldots, c_g \in \mathbb{C}$, not all zero, such that

$$c_0\theta_k(a) = \sum_{\nu=1}^{g} c_\nu \frac{\partial\theta_k}{\partial z_\nu}(a) \quad k = 0, \ldots, N .$$

We then have

$$c_0\big(\vartheta(a+u)\vartheta(a+v)\vartheta(a-u-v)\big) = \sum_{\nu=1}^{g} c_\nu \frac{\partial}{\partial z_\nu}\big(\vartheta(a+u)\vartheta(a+v)\vartheta(a-u-v)\big) \quad \forall u, v \in \mathbb{C}^g .$$

If we set $\varphi(z) = \left(\sum_{\nu=1}^{g} c_\nu \frac{\partial\vartheta}{\partial z_\nu}(z)\right) / \vartheta(z)$ this can be written

$$\varphi(a+u) + \varphi(a+v) + \varphi(a-u-v) = c_0 .$$

A priori, φ is meromorphic and has poles at the zeros Z of ϑ in $\mathbb{C}^g$. However, given a and $u_0 \in \mathbb{C}^g$, we can find a neighbourhood U of u_0 and $v \in \mathbb{C}^g$ such that $a+v \notin Z$ and $a-u-v \notin Z$ for $u \in U$. Thus φ is holomorphic on $\mathbb{C}^g$. Moreover, we have

$$\varphi(z+e_k) = \varphi(z) \quad \text{and} \quad \varphi(z+B_k) - \varphi(z) = \sum_{\nu=1}^{g} c_\nu \frac{\partial}{\partial z_\nu}(-2\pi i z_k - \pi i B_{kk}) = -2\pi i c_k .$$

It follows, as earlier, that φ is of the form

$$\varphi(z) = \alpha_0 + \alpha_1 z_1 + \cdots + \alpha_g z_g , \quad \alpha_0, \ldots, \alpha_g \in \mathbb{C} .$$

But φ is periodic, of period 1 in each z_j; hence $\alpha_j = 0$ for $j = 1, \ldots, g$, so that φ is constant. But then

$$-2\pi i c_k = \varphi(z+B_k) - \varphi(z) = 0 , \quad k = 1, \ldots, g ;$$

hence $\varphi(z) = \sum_{\nu=1}^{g} c_\nu \frac{\partial\vartheta}{\partial z_\nu}/\vartheta \equiv 0$, and $c_0 = \varphi(a+u) + \varphi(a+v) + \varphi(a-u-v) = 0$. This contradicts our assumption that not all the $c_0, \ldots, c_g$ are 0, and proves that $\theta : J(X) \to \mathbb{P}^N$ is an immersion.

This proves the theorem.

17. The Theta Divisor

In this section, we study the influence of the theta divisor on the Riemann surface X. The results were given by Riemann in his fundamental paper on abelian functions. The proofs given here are not very different from Riemann's.

Let L be the line bundle on $J(X)$ defined by the factor of automorphy $\varphi_{e_k} \equiv 1$, $\varphi_{B_k}(z) = e^{-2\pi i z_k - \pi i B_{kk}}$; the ϑ-function is a holomorphic section of L, and the theta divisor Θ is the divisor of the section ϑ. We denote by $\Theta_\zeta = \Theta + \zeta$ the translate of Θ by $\zeta \in J(X)$ [addition in $J(X)$]. It is defined by the section $\vartheta(z-\zeta)$ of the translate L_ζ of L by ζ.

Let $A : X \to J(X)$ be the Abel–Jacobi map; the function $P \mapsto \vartheta(A(P)-\zeta)$ is a section of the pull-back $A^*(L_\zeta)$ of L_ζ by A. If we choose a basis a_j, b_j of $H_1(X,\mathbb{Z})$ as in §14 and slit X along these curves, we obtain a polygon (simply connected) $\Delta \subset \mathbb{C}$ and $\vartheta(A(P)-\zeta)$ may be thought of as a holomorphic function on Δ. Note that the a_j, b_j can be chosen to avoid any given finite subset of X; in what follows, we shall tacitly assume that this has been done. The sides of the $4g$ – gon Δ will be denoted, as in §14, by $a_\nu, b_\nu, a'_\nu, b'_\nu$ (a'_ν, b'_ν map onto the curves a_ν^{-1}, b_ν^{-1} in X).

Theorem 1. *Let $\zeta \in J(X)$ be such that $A(X) \not\subset \Theta_\zeta$ [the set $\{\zeta \in J(X) | A(X) \subset \Theta_\zeta\}$ is clearly a proper analytic subset of $J(X)$]. Then, counted with multiplicities, the intersection $A(X) \cap \Theta_\zeta$ consists of g points; more precisely, the divisor of the section $\vartheta(A(P)-\zeta)$ of $A^*(L_\zeta)$ has degree g: $(\vartheta(A(P)-\zeta)) = \sum_{i=1}^{g} P_i(\zeta)$.*

Further, we have

$$\sum_{i=1}^{g} A(P_i(\zeta)) = \zeta - \kappa$$

where $\kappa \in J(X)$ is a point independent of ζ (it depends only on the base point $P_0 \in X$ chosen to define the Abel–Jacobi map).

Proof. Let $\vec{\omega} = (\omega_1, \ldots, \omega_g)$, where $\omega_1, \ldots, \omega_g$ is a normalized basis of $H^0(X,\Omega)$: $\int_{a_\nu} \omega_k = \delta_{\nu k}$. On Δ, the Abel-Jacobi map is given, modulo Λ, by

$$A(P) = (A_1(P), \ldots, A_g(P)) = \int_{P_0}^{P} \vec{\omega}\,.$$

If Φ is a function on $\partial\Delta$, we define functions $\Phi^\pm$ on the edges a_j, b_j of $\partial\Delta$ by $\Phi^+ = \Phi$, $\Phi^-(P) = \Phi(P')$ if $P \in a_j$ or b_j and P' is the corresponding point of a'_j, b'_j.

If $P \in a_\nu$, we have, as in §14, Lemma 1, (see figure next page)

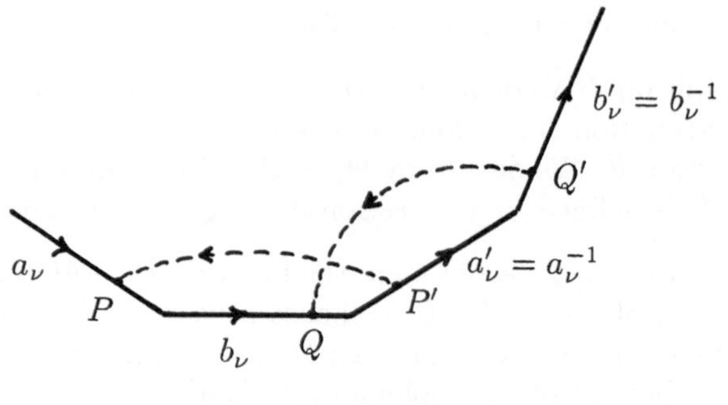

$A_k^+(P) - A_k^-(P) = \int_{P'}^P \omega_k = -\int_{b_\nu} \omega_k = -B_{\nu k}$, while if $Q \in b_\nu$, we have $A_k^+(Q) - A_k^-(Q) = \int_{Q'}^Q \omega_k = \int_{a_\nu} \omega_k = \delta_{\nu k}$. Thus, if $A^\pm = (A_1^\pm, \dots, A_g^\pm)$, we have

$$A^+ - A^- = e_\nu \quad \text{on} \quad b_\nu\,, \quad A^+ - A^- = -B_\nu \quad \text{on} \quad a_\nu\,.$$

We may assume that $\vartheta\big(A(P) - \zeta\big) \neq 0$ if $P \in \partial\Delta$. The number of zeros of $F(P) = \vartheta\big(A(P) - \zeta\big)$ in Δ is given by

$$\frac{1}{2\pi i}\int_{\partial\Delta} d\log F(P) = \frac{1}{2\pi i}\sum_{\nu=1}^{g}\Big(\int_{a_\nu} + \int_{b_\nu}\Big) d\log\frac{F^+(P)}{F^-(P)}\,.$$

Now, if $P \in b_\nu$, $F^+(P) = \vartheta\big(A^+(P) - \zeta\big) = \vartheta\big(A^-(P) - \zeta + e_\nu\big) = F^-(P)$, while, if $P \in a_\nu$, $F^+(P) = \vartheta\big(A^-(P) - \zeta - B_\nu\big) = e^{2\pi i\left(A_\nu(P) - \zeta_\nu\right) + \pi i B_{\nu\nu}}\vartheta\big(A^-(P) - \zeta\big)$ so that $\log\frac{F^+(P)}{F^-(P)} = 2\pi i A_\nu(P) - 2\pi i\zeta_\nu + \pi i B_{\nu\nu}$, and we have $d\log\frac{F^+}{F^-} = 2\pi i\omega_\nu$ on a_ν. Hence, the number of zeros of F in Δ equals $\sum_{\nu=1}^g \int_{a_\nu} \omega_\nu = g$. This proves the first part of the theorem.

For the second part, let $P_1(\zeta), \dots, P_g(\zeta)$ be the zeros of $\vartheta\big(A(P) - \zeta\big)$ in Δ. We shall denote by *const* a term which is independent of ζ. We have

$$\begin{aligned}\sum_{\nu=1}^{g} A_k\big(P_\nu(\zeta)\big) &= \frac{1}{2\pi i}\int_{\partial\Delta} A_k(P)\, d\log F(P) \\ &= \frac{1}{2\pi i}\sum_{\nu=1}^{g}\Big(\int_{a_\nu} + \int_{b_\nu}\Big)\big(A_k^+ d\log F^+ - A_k^- d\log F^-\big)\,.\end{aligned}$$

Consider the integral over a_ν. We have $A_k^- = A_k^+ + B_{\nu k}$, while $d\log F^+ = d\log F^- + 2\pi i\omega_\nu$; hence

$$\begin{aligned}\int_{a_\nu}\big(A_k^+ d\log F^+ - A_k^- d\log F^-\big) &= \int_{a_\nu}(A_k^+ - A_k^-)\, d\log F^+ + 2\pi i\int_{a_\nu} A_k^-\omega_\nu \\ &= -B_{\nu k}\int_{a_\nu} d\log F^+ + \mathit{const}\,.\end{aligned}$$

If α, β are the ordered extremities of a_ν, we have $A^+(\beta) - A^+(\alpha) = \int_{a_\nu} \vec{\omega} = e_\nu$. Hence

$$\begin{aligned} \frac{1}{2\pi i}\int_{a_\nu} d\log F^+ &\equiv \frac{1}{2\pi i}\log\frac{\vartheta(A^+(\beta)-\zeta)}{\vartheta(A^+(\alpha)-\zeta)} \mod \mathbb{Z} \\ &\equiv \frac{1}{2\pi i}\log\frac{\vartheta(A^+(\alpha)-\zeta+e_\nu)}{\vartheta(A^+(\alpha)-\zeta)} \equiv 0 \mod \mathbb{Z}\,. \end{aligned}$$

Since the integral depends continuously on ζ, we conclude that

$$\int_{a_\nu} \left(A_k^+ d\log F^+ - A_k^- d\log F^-\right) = \mathit{const} \text{ for } \nu = 1, \ldots, g\,.$$

Consider the integral over b_ν. We have $A^+ = A^- + e_\nu$, $F^+ = F^-$ on b_ν, so that

$$\int_{b_\nu} \left(A_k^+ d\log F^+ - A_k^- d\log F^-\right) = \delta_{\nu k}\int_{b_\nu} d\log F^+\,.$$

If x, y denote the ordered endpoints of b_ν, we have $A(y) = A(x) + B_\nu$, so that $\frac{F^+(y)}{F^+(x)} = \frac{\vartheta\left(A(x)-\zeta+B_\nu\right)}{\vartheta\left(A(x)-\zeta\right)} = \exp\left(-2\pi i A_\nu(x) + 2\pi i\zeta_\nu - \pi i B_{\nu\nu}\right)$, so that

$$\frac{1}{2\pi i}\int_{b_\nu} d\log F^+ \equiv \zeta_\nu - A_\nu(x) - \frac{1}{2}B_{\nu\nu} \mod \mathbb{Z}$$

and we conclude that $\frac{1}{2\pi i}\int_{b_\nu} d\log F^+ = \zeta_\nu + \mathit{const}$. This gives

$$\sum_{\nu=1}^{g} A_k(P_\nu(\zeta)) = \sum_{\nu=1}^{g} \delta_{\nu k}\zeta_\nu + \mathit{const} = \zeta_k + \mathit{const}$$

which proves the theorem.

Before proceeding to the next theorem, we need some preliminaries.

If $0 < k \leq g$, and $S^k(X)$ is the k-th symmetric power of X, we denote by W_k the image in $J(X)$ of the map $A : S^k(X) \to J(X)$ $[A(P_1, \ldots, P_k) = \Sigma A(P_i)]$. W_k is thus the set $W_k = \{A(D) | D$ effective divisor on X of degree $k\}$. W_k is an analytic set in $J(X)$.

Let M be a compact connected complex manifold with $\dim_{\mathbb{C}} M = n$. A *divisor* D on M is a finite linear combination $D = \sum_{k=1}^{n} n_k Y_k$, $b_k \in \mathbb{Z}$, where the Y_k are irreducible analytic subsets of M of dimension $n-1$ (i.e. codimension 1). On a complex manifold M, codimension 1 analytic sets $Y \subset M$ have *local equations*, i.e. $\forall a \in M$, there is a neighbourhood U of a and f holomorphic on U such that, if $x \in U$ and g is a holomorphic function near x vanishing on Y near x, then g is a multiple of f by a holomorphic function near x.

If $U \subset M$ and f_k are local equations for Y_k on U, set $f_U = \prod f_k^{n_k}$. If V is another such open set and f_V the corresponding meromorphic function on V, then $f_U = g_{UV} f_V$ on $U \cap V$, where g_{UV} is holomorphic and nowhere zero on $U \cap V$. The $\{g_{UV}\}$ form transition functions for a holomorphic line bundle $L = L(D)$ on M. It comes with a standard section (meromorphic) s_D defined by the function $\{f_U\}$.

If $D = \Sigma n_k Y_k$ is a divisor, the set $\bigcup_{n_k \neq 0} Y_k$ is called the support of D, and written supp(D). If $n_k \geq 0$ for all k, D is called effective. The standard section s_D is holomorphic exactly when D is effective.

Meromorphic sections s of a holomorphic line bundle L define divisors on M. If Y is the analytic set of zeros and poles of s, and $Y = \bigcup Y_k$ its decomposition into irreducible components, let s be represented locally by a meromorphic function F and f_k be local equations for Y_k. Then $F = u \cdot \prod f_k^{n_k}$, u a holomorphic function without zeros. The n_k are constant along Y_k and we set $(s) = \Sigma n_k Y_k$. The integer n_k is called the order of s_k along Y_k [order of zero or pole according as $n_k > 0$ or $n_k < 0$; the order of the pole is $|n_k|$ if $n_k < 0$].

Theorem 2. *We have*

$$\Theta = W_{g-1} + \kappa\,,$$

where κ is the constant in Theorem 1.

In other words, the divisor Θ has the form $1 \cdot Y$ where Y is irreducible of dimension $g-1$ (so that the theta function has only simple zeros at a general point of Θ); moreover, Θ consists exactly of the points $\sum_{\nu=1}^{g-1} A(P_\nu) + \kappa$, $P_1, \ldots, P_{g-1} \in X$.

Proof. We start by showing that $W_{g-1} + \kappa \subset \text{supp}(\Theta)$. Let $D = P_1 + \cdots + P_g$ be a divisor of degree g with distinct P_i in general position so that D is the unique point of $S^g(X)$ mapping onto $A(D)$ in $J(X)$. Further, we may assume that $A(X) \not\subset \Theta_\zeta$, where $\zeta = A(D) + \kappa$ (since $A : S^g(X) \to J(X)$ is surjective).

Let $Q_1, \ldots, Q_g$ be the zeros of $P \mapsto \vartheta\big(A(P) - \zeta\big)$. By Theorem 1 we have $\Sigma A(Q_i) = \zeta - \kappa = A(D)$, so that, by choice of D, we have $D = \Sigma Q_i = \Sigma P_\nu$. In particular $\vartheta\big(A(P_g) - \zeta\big) = 0$, so that $0 = \vartheta\big(-\sum_{\nu=1}^{g-1} A(P_\nu) - \kappa\big) = \vartheta\big(\sum_{\nu=1}^{g-1} A(P_\nu) + \kappa\big)$. Since D can be chosen to satisfy the above conditions arbitrarily in a non-empty open set in $S^g(X)$, it follows that $\vartheta\big(A(D') + \kappa\big) = 0$ for all D' in a non-empty open set in $S^{g-1}(X)$, so that $\vartheta | W_{g-1} + \kappa = 0$.

To prove that $\text{supp}(\Theta) \subset W_{g-1} + \kappa$, let $\zeta \in \Theta$, and suppose first that there is $P \in X$ such that

$$\vartheta\big(A(x) - A(P) - \zeta\big) \not\equiv 0 \text{ in } x\,;$$

in this case, if $D = \text{div}\big(\vartheta\big(A(x) - A(P) - \zeta\big)\big)$, then $D = P + D'$ where $D' \geq 0$ has degree $g-1$ ($P \in \text{supp}(D)$ since $\vartheta(-\zeta) = \vartheta(\zeta) = 0$); moreover, by Theorem 1,

$$A(D) = A(P) + A(D') = \big(\zeta + A(P)\big) - \kappa\,,$$

so that $\zeta = A(D') + \kappa \in W_{g-1} + \kappa$.

If $\vartheta(A(x) - A(P) - \zeta) \equiv 0 \ \forall P$, let k be the largest integer such that $\vartheta(A(D_0) - A(D_1) - \zeta) = 0$ for all effective divisors D_0, D_1 of degree k. We have $k < g$ since $S^g(X) \to J(X)$ is surjective.

Let E_0, E_1 be effective divisors of degree $k+1$ with $\vartheta(A(E_0) - A(E_1) - \zeta) \neq 0$. We may suppose that $\operatorname{supp}(E_0 + E_1)$ consists of $2k+2$ distinct points. Let $E_0 = P + D_0$ where $D_0 \geq 0$ has degree k. Then, $x \mapsto \vartheta(A(x) + A(D_0) - A(E_1) - \zeta) \not\equiv 0$ (it is $\neq 0$ for $x = P$); let D be the divisor of this function. Then $D \geq 0$ has degree g. Further, if $x \in \operatorname{supp}(E_1)$, $\vartheta(A(x) + A(D_0) - A(E_1) - \zeta) = \vartheta(A(D_0) - A(E_1 - x) - \zeta) = 0$ since $E_1 - x \geq 0$ has degree k. Hence $D \geq E_1$, and we can write $D = E_1 + E_2$ with $\deg(E_2) = g - k - 1$.

Now, by Theorem 1, $A(E_1) + A(E_2) = A(D) = \zeta + A(E_1) - A(D_0) - \kappa$, so that $\zeta - \kappa = A(E_2 + D_0)$ with $\deg(E_2 + D_0) = g - k - 1 + k = g - 1$. Thus, $\operatorname{supp}(\Theta) \subset W_{g-1} + \kappa$.

We saw at the beginning of this proof that if $\zeta = A(D) + \kappa$, with $D = \Sigma P_i$ with the P_i distinct in general position, then D is the divisor of zeros of $\vartheta(A(x) - \zeta)$, $\zeta = A(D) + \kappa$, so that the *zeros of* $\vartheta(A(x) - \zeta)$ *are simple.* It follows that

$$\Theta = W_{g-1} + \kappa \quad \text{as divisors} .$$

Theorem 3. *If κ is the constant in Theorems 1 and 2, and K_X is a canonical divisor on X, we have*

$$A(K_X) = -2\kappa .$$

Proof. We begin with a remark which we shall use later on in these notes as well.

Let $D \geq 0$ be a divisor of degree $g-1$. Then $h^0(D) \geq 1$. By the Riemann–Roch theorem $h^0(K_X - D) = h^0(D) - (1 - g + \deg D) = h^0(D) \geq 1$, so that $K_X - D$ is linearly equivalent to a divisor $D' \geq 0$ which must have degree $g-1$. Hence $A(K_X - D) \in W_{g-1}$. Hence $A(K_X) - W_{g-1} \subset W_{g-1}$. Further, $A(D) = A(K_X) - A(D') \in A(K_X) - W_{g-1}$.

Thus, we have

$$A(K_X) - W_{g-1} = W_{g-1} \, .$$

Since $\vartheta(z) = \vartheta(-z)$, we have

$$\Theta = W_{g-1} + \kappa = -\Theta = -W_{g-1} - \kappa = W_{g-1} - A(K_X) - \kappa = \Theta - (A(K_X) + 2\kappa) \, .$$

Since Θ is not left invariant by translation by a non-zero element of $J(X)$ (§16, Lemma 4), we obtain

$$A(K_X) + 2\kappa = 0 \, .$$

Theorem 4. *Let $\zeta \in J(X)$. Then $A(X) \subset \Theta_\zeta$ if and only if $\zeta - \kappa = A(D)$ where $D \geq 0$ is an effective divisor of degree g with $\dim |D| > 0$; in other words D is a special divisor of degree g.*

Proof. $A(X) \subset \Theta_\zeta$ if and only if $A(P) - \zeta \in \Theta \ \forall P \in X$, i.e. $\zeta - A(P) \in \Theta = W_{g-1} + \kappa \ \forall P \in X$. Thus, the condition is that $\zeta - \kappa = A(D)$ where D has degree g and $P \in \operatorname{supp}(D)$. Now, D is determined up to linear equivalence; if we fix D_0 with $A(D_0) = \zeta - \kappa$, the condition is that there is $D \geq 0$ linearly equivalent to D_0 and containing an arbitrarily given point $P \in X$. This simply means that $\dim |D_0| > 0$.

Corollary. If $\zeta \in J(X)$ is such that $A(X) \not\subset \Theta_\zeta$, then there is a unique divisor $D \geq 0$ of degree g such that $A(D) + \kappa = \zeta$. D is given by the divisor of zeros of $\vartheta(A(P) - \zeta)$.

This follows from Theorem 4 and §15, Theorems 1,2.

This corollary gives a complete answer to the so-called *Jacobi inversion problem*, viz to describe the inverse of the birational transformation $A : S^g(X) \to J(X)$.

We shall give another application of these results. Consider the map $A : S^g(X) \to J(X)$, $(P_1, \dots, P_g) \mapsto \Sigma A(P_i)$, and let $Y \subset S^g(X)$ be the set of critical points, i.e. $Y = \{D \in S^g(X) \mid \operatorname{rank}_D(dA) < g\}$. Y is an analytic set of dimension $\leq g - 1$. Further, if $D \in Y$, then $A^{-1}A(D) \subset Y$ (by §15, Theorem 2 and Abel's theorem) and the dimension of $A^{-1}A(D)$ at any of its points is $\dim |D| > 0$. Hence $Y' = A(Y)$ is an analytic set in $J(X)$ of dimension $\leq g - 2$. In particular, *no finite union of translates of Y' can contain Θ.*

Let now $P \in X$ and let x be a variable point on X. By Theorem 4, if $A(P) + \zeta - \kappa \notin Y'$, then the function $x \mapsto \vartheta(A(x) - A(P) - \zeta)$ has exactly g zeros $P_1, \dots, P_g$, and $\Sigma A(P_i) = \zeta + A(P) - \kappa$; further ΣP_i is the only divisor ≥ 0 of degree g satisfying this equation.

If we assume, in addition that $\zeta \in \Theta$, $\zeta = A(Q_1^0) + \dots + A(Q_{g-1}^0) + \kappa$, then

$$\sum_1^g A(P_i) = A(P) + \sum_{j=1}^{g-1} A(Q_j^0) ,$$

so that $\Sigma P_i = P + \Sigma Q_j^0$.

Thus, if $\zeta \in \Theta$ and $\zeta \notin -A(P) + \kappa + Y'$, *then the zeros of $x \mapsto \vartheta(A(x) - A(P) - \zeta)$ are given by*

$$(P, Q_1^0, \dots, Q_{g-1}^0) ,$$

where $Q_1^0, \dots, Q_{g-1}^0$ depend only on ζ and not on P.

Consider now a non-constant meromorphic function f on X, and let $(f) = \sum_{k=1}^r P_k - \sum_{k=1}^r Q_k$. We choose $\zeta \in \Theta$, $\zeta \notin \bigcup_k (Y' + \kappa - A(P_k)) \cup \bigcup_k (Y' + \kappa - A(Q_k))$, and write $\zeta = A(D_0) + \kappa$, $D_0 = A(Q_1^0) + \dots + A(Q_{g-1}^0)$. We transform X into a simply connected polygon Δ as in §14, with the a_ν, b_ν avoiding a suitable finite set of points in X. Consider the function on Δ defined by

$$F(x) = \prod_{k=1}^{r} \frac{\vartheta\big(A(x) - A(P_k) - \zeta\big)}{\vartheta\big(A(x) - A(Q_k) - \zeta\big)}\,.$$

Its divisor $= (\Sigma P_k + rD_0) - (\Sigma Q_k + rD_0) = \Sigma(P_k - Q_k) = (f)$. It is not, however, a function defined on X.

If $x \in b_\nu$ and x' is the corresponding point of b'_ν, then $A(x') = A(x) + e_\nu$, and we have $F(x) = F(x')$.

If $x \in a_\nu$ and x' is the corresponding point of a'_ν, we have $A(x') = A(x) + B_\nu$ and

$$\frac{F(x')}{F(x)} = \frac{\prod_{k=1}^{r} e^{-2\pi i\left(A_\nu(x) - A_\nu(P_k) - \zeta_\nu\right)}}{\prod_{k=1}^{r} e^{-2\pi i\left(A_\nu(x) - A_\nu(Q_\nu) - \zeta_\nu\right)}} = \exp\Big\{2\pi i \sum_{1}^{r} \big(A_\nu(P_k) - A_\nu(Q_k)\big)\Big\}\,.$$

By Abel's theorem, there exist integers $n_1, \ldots, n_g$, $m_1, \ldots, m_g$ so that

$$\begin{aligned}\sum_{1}^{r} \big(A_\nu(P_k) - A_\nu(Q_k)\big) &= \nu^{\text{th}} \text{ component of } \sum_{j=1}^{g} n_j e_j + \sum_{j=1}^{g} m_j B_j \\ &= n_\nu + \sum_{1}^{g} m_j B_{j\nu}\,.\end{aligned}$$

Now if $\omega_1, \ldots, \omega_g$ is the normalized basis of $H^0(X, \Omega)$, and $\omega = \sum_{j=1}^{g} m_j \omega_j$, and we set $\varphi(x) = \int_{P_0}^{x} \omega$ (P_0 fixed), we find that $e^{2\pi i \varphi(x)}$ has the two properties:

1) If $x \in b_\nu$ and x' is the corresponding point of b'_ν, then $e^{2\pi i \varphi(x)} = e^{2\pi i \varphi(x')}$.

2) If $x \in a_\nu$ and x' is the corresponding point of a'_ν, then $e^{2\pi i \varphi(x')} \,/\, e^{2\pi i \varphi(x)} = e^{2\pi i \sum_j m_j B_{j\nu}}$. Thus $F(x) e^{-2\pi i \varphi(x)}$ defines a meromorphic function on X with divisor $\Sigma(P_k - Q_k) = (f)$. Thus, we have proved

Theorem 5. (Riemann's Factorisation Theorem). *Let f be a non-constant meromorphic function on X, with divisor $(f) = \sum_{k=1}^{r} P_k - \sum_{k=1}^{r} Q_k$. Then, there exists $\omega \in H^0(X, \Omega)$ such that, if ζ is a general point on Θ, we have*

$$f(x) = c \cdot e^{\int_{P_0}^{x} \omega} \prod_{k=1}^{r} \frac{\vartheta\big(A(x) - A(P_k) - \zeta\big)}{\vartheta\big(A(x) - A(Q_k) - \zeta\big)}\,.$$

This theorem provides the analogue of the factorisation of rational functions into linear factors.

In the same way, one can prove

Theorem 6. *Let $P, Q \in X$, $P \neq Q$. Then, if ζ is a general point of Θ,*

$$d_x \log \frac{\vartheta(A(x) - A(P) - \zeta)}{\vartheta(A(x) - A(Q) - \zeta)}$$

is a meromorphic 1-form on X, holomorphic on $X - \{P, Q\}$, and with simple poles at P, Q, with residue $+1$ at P and -1 at Q. Moreover, its a-periods are 0 since ϑ is periodic.

One can also construct forms with higher order poles at one point, having residue 0, in the same way.

The idea behind Riemann's factorisation theorem can be used to construct functions with certain essential singularities on X. These functions are of great importance in the study of certain non-linear partial differential equations which have turned out to be very closely connected with the geometry of algebraic curves. For an introduction to this circle of ideas, one may consult

B.A. Dubrovin: Theta functions and non-linear equations *Russian Math. Surveys* (Uspekhi) 36 (1981), 11 – 92.

I.M. Krichever and S.P. Novikov: Holomorphic bundles over algebraic curves and non-linear equations, *Russian Math. Surveys* (Uspekhi) 35 (1980), 53 – 79.

D. Mumford: *Tata Lectures on Theta*, 2 vols., Birkhäuser, 1983, 1984.

T. Shiota: Characterization of Jacobian varieties in terms of soliton equations, *Inventiones Math.* 83 (1986), 333 – 382.

The literature surrounding this relationship between algebraic curves and non-linear PDE has become enormous.

Let P be a fixed point on the compact Riemann surface X, let (U, z) be a local coordinate at P, $z(P) = 0$, and let u be a polynomial in 1-variable over $\mathbb{C}$. Let D be a non-special effective divisor of degree g on X. We assume that $P \notin \mathrm{supp}(D)$.

Theorem 7. *There exists a function F meromorphic on $X - P$ such that*

(i) $(F) \geq -D$ *on $X - P$;*

(ii) $F(z)\exp(-2\pi i u(\frac{1}{z}))$ *is holomorphic at P.*

Proof. Let $u(t) = c_0 + c_1 t + \cdots + c_r t^r$ $(c_r \neq 0)$. Let $\omega_P^{(n)}$ be the normalized abelian differential of the second kind with pole $\frac{dz}{z^{n+1}}$ $(n \geq 1)$ at P (holomorphic on $X - P$). Then $du(\frac{1}{z}) + \sum_{n=1}^{r} n c_n \omega_P^{(n)}$ is holomorphic at P. Let $\varphi = -\sum_{n=1}^{r} n c_n \omega_P^{(n)}$. Let $\beta = (\beta_1, \ldots, \beta_g)$ be the vector of b-periods of φ:

$$\beta_\nu = \int_{b_\nu} \varphi\,;$$

(we have $\int_{a_\nu}\varphi = 0$ since the $\omega_P^{(n)}$ are normalized).

Let

$$F(x) = \exp\Big(2\pi i \int_{P_0}^{x}\varphi\Big)\frac{\vartheta\big(A(x)-A(D)+\beta-\kappa\big)}{\vartheta\big(A(x)-A(D)-\kappa\big)}\,.$$

First, F is single valued on X: consider it as a function on Δ as in Theorem 5. If $x \in b_\nu$ and x' is the corresponding point of b'_ν, then $A(x') = A(x) + e_\nu$ and $F(x) = F(x')$ since $\int_{x'}^{x}\varphi = \int_{a_\nu}\varphi = 0$.

If $x \in a_\nu$ and x' is the corresponding point of a'_ν, then

$$\int_{x'}^{x}\varphi = -\int_{b_\nu}\varphi = -\beta_\nu\,,$$

while $A(x) - A(x') = -B_\nu$, so that $\vartheta\big(A(x)-A(D)+\beta-\kappa\big) = \vartheta\big(A(x')-A(D)+\beta-\kappa\big)\times\exp\big(2\pi i\big(A_\nu(x')-A_\nu(D)+\beta_\nu-\kappa_\nu\big)+\pi i B_{\nu\nu}\big)$, and $\vartheta\big(A(x)-A(D)-\kappa\big) = \vartheta\big(A(x')-A(D)-\kappa\big)\exp\big(2\pi i\big(A(x')-A_\nu(D)-\kappa_\nu\big)+\pi i B_{\nu\nu}\big)$. Hence

$$\frac{F(x)}{F(x')} = e^{-2\pi i\beta_\nu}\cdot\frac{\exp\big(2\pi i\big(A_\nu(x')-A_\nu(D)+\beta_\nu-\kappa_\nu\big)+\pi i B_{\nu\nu}\big)}{\exp\big(2\pi i\big(A_\nu(x')-A_\nu(D)-\kappa_\nu\big)+\pi i B_{\nu\nu}\big)} = 1\,.$$

Since $\varphi - du(\frac{1}{z})$ is holomorphic at P, $Fe^{-2\pi i u(\frac{1}{z})}$ is holomorphic at P.

Finally, the poles of F are at the zeros of $\vartheta\big(A(x)-A(D)-\kappa\big)$ and $\operatorname{div}\big(\vartheta\big(A(x)-A(D)-\kappa\big)\big) = D$ since D is non-special.

It is not hard to see that if D and u are generic, this function is uniquely determined up to a constant multiple. In fact, one sees that the divisor D' of zeros of the function constructed above is non-special; if F_0 is a function satisfying the conditions of Theorem 7, then F_0/F is meromorphic on X, and $(F_0/F) \geq -D'$; since D' is non-special, F_0/F is constant.

18. Torelli's Theorem

Torelli's theorem asserts that the pair $(J(X), \Theta)$ determines the Riemann surface X. There are several proofs available of this theorem. The one we shall give here is due to Henrik Martens [12]. There are more "geometric" proofs, some of which will be found in Griffiths–Harris [9] or Arbarello–Cornalba–Griffiths–Harris [10]. We begin with a general fact about complex tori.

Let $T_1 = \mathbb{C}^m/\Lambda_1$, $T_2 = \mathbb{C}^n/\Lambda_2$ be two complex tori (Λ_1 is a lattice in $\mathbb{C}^m$, Λ_2, a lattice in $\mathbb{C}^n$).

Lemma 1. *Let $f : T_1 \to T_2$ be a holomorphic map and $F : \mathbb{C}^m \to \mathbb{C}^n$ be a lifting of f. Then F is a polynomial of degree ≤ 1.*

Proof. Since F lifts f, if $\lambda \in \Lambda_1$, then $F(z+\lambda) - F(z) \in \Lambda_2$ for all $z \in \mathbb{C}^m$, and so is constant. Hence, for $1 \leq \nu \leq m$, $\frac{\partial F}{\partial z_\nu}$ is invariant under translation by λ, so defines a holomorphic map $T_1 \to \mathbb{C}^n$, which is constant since T_1 is compact. This proves the lemma.

TORELLI'S THEOREM. *Let X, Y be compact Riemann surfaces of genus $g \geq 1$, and Θ_X, Θ_Y the theta divisors on $J(X)$, $J(Y)$ respectively.*

Suppose that there exists an analytic isomorphism $\varphi : J(X) \to J(Y)$ such that $\varphi^*(\Theta_Y) = \Theta_X$. Then X and Y are analytically isomorphic.

Proof. We shall assume that $(J(X), \Theta_X)$ and $(J(Y), \Theta_Y)$ have been identified by φ.

We denote by $A_X : X \to J(X)$, the Abel-Jacobi map of X, and by W_r the image in $J(X)$ of $S^r(X)$, $1 \leq r \leq g$

We let $A_Y : Y \to J(Y)$ be the Abel-Jacobi map of Y, and set V_r = image of $S^r(Y)$ in $J(Y)$, $1 \leq r \leq g$.

We shall actually prove the following theorem, which clearly implies Torelli's theorem

Theorem. *Assume that W_{g-1} is a translate of V_{g-1}. Then V_1 is a translate of either W_1 of $-W_1$.*

We start with some notation.

If E is a subset of $J(X)$, we denote by E^* the set $E^* = A(K_X) - E$, where K_X is a canonical divisor on X. We shall call E^* the dual of E.

We have (see proof of §17, Theorem 3)

$$W_{g-1}^* = W_{g-1}\ .$$

For any set $E \subset J(X)$ and $a \in J(X)$, we denote by E_a the translate of E by a: $E_a = E + a$. With this notation, we have

$$(W_{g-1,a})^* = W_{g-1,-a}\ .$$

We shall also denote an effective divisor of degree k on X by $D_k, D'_k, \Delta_k, \ldots$ (In other words, subscripts on a divisor will indicate its degree in the proof of the above theorem).

Lemma 1. *Let* $0 \le r \le g-1$, $a, b \in J(X)$. *Then* $W_{r,a} \subset W_{g-1,b}$ *if and only if* $a \in W_{g-1-r,b}$.

Proof. If $a = A_X(D_{g-1-r}) + b$, then $A(D_r) + a = A(D_r + D_{g-1-r}) + b \in W_{g-1,b}$.

To prove the converse, we may assume that $b = 0$. By assumption, $\forall\ D_r$ ($D_r \ge 0$ of degree r), there is Δ_{g-1} such that $A_X(D_r) + a = A_X(\Delta_{g-1})$. Now, if P_0 is the base point in X defining the Abel–Jacobi map, we have $A_X(rP_0) = 0$, so that $a = A_X(\delta)$, where $\delta \ge 0$ has degree $g-1$. We now have $A_X(D_r + \delta) = A_X(\Delta_{g-1} + rP_0)$, so that, by Abel's theorem $D_r + \delta \sim \Delta_{g-1} + rP_0$ (linear equivalence). Hence $D_r + K_X - \Delta_{g-1} \sim (K_X - \delta) + rP_0$; moreover $K_X - \Delta_{g-1}$ and $K_X - \delta$ are linearly equivalent to effective divisors [since $h^0(D'_{g-1}) = h^0(K_X - D'_{g-1})$]. Thus, $K_X - \delta + rP_0$ is linearly equivalent to a divisor of the form $D_r + D'_{g-1}$ $\forall D_r$; hence $\dim |K_X - \delta + rP_0| \ge r$. Hence, by the Riemann–Roch theorem $h^0(\delta - rP_0) = h^0(K_X - \delta + rP_0) + 1 - g + (g-1-r) \ge 1$, so that $\delta - rP_0 \sim D^0_{g-1-r}$, and we have $A_X(D^0_{g-1-r}) = A_X(\delta - rP_0) = A_X(\delta) = a$, and $a \in W_{g-1-r}$.

Lemma 2. *Let* $0 \le r \le g-1$. *We have*

$$W_{g-1-r} = \bigcap_{a \in W_r} W_{g-1,-a}\ ,$$

and

$$W_{g-1-r}^* = \bigcap_{a \in W_r} W_{g-1,a} = \bigcap_{a \in W_r} (W_{g-1,-a})^*\ .$$

Proof. If $a \in W_r$, we have $a = A_X(D_r)$ and $W_{g-1-r} + A_X(D_r) \subset W_{g-1}$, so that $W_{g-1-r} \subset \bigcap_{a \in W_r} W_{g-1,-a}$.

Let now $\zeta \in \bigcap_{a \in W_r} W_{g-1,-a}$, so that $\zeta + W_r \subset W_{g-1}$. By Lemma 1, this implies that $\zeta \in W_{g-1-r}$, and the first statement is proved. The second follows from the first by taking duals.

Lemma 3. *Let* $0 \le r \le g-2$, *let* $a \in J(X)$, $x \in W_1$, $y \in W_{g-1-r}$. *Set* $b = a + x - y$. *Then we have: Either*

$$W_{r+1,a} \subset W_{g-1,b}$$

or

$$W_{g-1,b} \cap W_{r+1,a} = W_{r,a+x} \cup S$$

where $S = W_{r+1,a} \cap (W_{g-2,y-a})^*$.

Proof. By definition of W_k, there is $P \in X$ with $A_X(P) = x$, and D^0_{g-1-r} such that $A_X(D^0_{g-1-r}) = y$.

1) Suppose that $P \in \operatorname{supp}(D^0_{g-1-r})$. Then $x - y = -A_X(D')$ where $\deg D' = g-2-r$ (and $D' \ge 0$), and we have

$$a = b + A_X(D') .$$

so that $a + W_{r+1} = b + \big(A_X(D') + W_{r+1}\big) \subset b + W_{g-1}$, which is our first alternative.

2) Suppose that $P \notin \operatorname{supp}(D^0_{g-1-r})$, and let

$$u \in W_{r+1,a} \cap W_{g-1,b} .$$

Then

$$u = A_X(D_{r+1}) + a = A_X(\Delta_{g-1}) + b = A_X(\Delta_{g-1}) + a + A_X(P) - A_X(D^0_{g-1})$$

so that, since $D_{r+1} + D^0_{g-1-r}$ and $\Delta_{g-1} + P$ both have degree g, Abel's theorem implies that

$$D_{r+1} + D^0_{g-1-r} \sim \Delta_{g-1} + P .$$

Case (i). $D_{r+1} + D^0_{g-1-r} = \Delta_{g-1} + P$.
Since $P \notin \operatorname{supp}(D^0_{g-1-r})$, we have $P \in \operatorname{supp}(D_{r+1})$, and we have

$$D'_r + D^0_{g-1-r} = \Delta_{g-1} , \quad (D'_r = D_{r+1} - P)$$

so that

$$A_X(D'_r) + y = u - b$$

and $u \in W_r + b + y = W_{r,a+x}$.

Case (ii). $D_{r+1} + D^0_{g-1-r} \ne \Delta_{g-1} + P$.
In this case, the complete linear system $|\Delta_{g-1} + P|$ contains two distinct effective divisors, so that $\dim |\Delta_{g-1} + P| \ge 1$. Hence, for *any* $Q \in X$, we can find a $\Delta'_{g-1} \ge 0$ so that $\Delta_{g-1} + P \sim \Delta'_{g-1} + Q$. This gives, if $w = A(Q)$, $(u-b) + x = A_X(\Delta_{g-1}) + A_X(P) = A_X(\Delta'_{g-1}) + w \in W_{g-1,w}$; since $Q \in X$ is arbitrary, $u - b + x \in \bigcap_{w \in W_1} W_{g-1,w} = W^*_{g-2}$

(by Lemma 1); thus $u \in (W^*_{g-2})_{b-x} = (W^*_{g-2})_{a-y} = (W_{g-2,y-a})^*$; of course, $u \in W_{r+1,a}$ by assumption. Thus, we have $W_{r+1,a} \cap W_{g-1,b} \subset W_{r,a+x} \cup S$.

We have still to check the opposite inclusion. We have $W_r + a + x = W_r + a + A_X(P) \subset W_{r+1,a}$. Since $a+x = b+y \in b+W_{g-1-r}$, we have $W_{r,a+x} \subset b+W_{g-1-r}+W_r = b+W_{g-1}$. Finally, $(W_{g-2,y-a})^* = W^*_{g-2}+b-x = A_X(K_X)-W_{g-2}-x+b \subset A_X(K_X)-W_{g-1}+b = W_{g-1}+b$.

This proves that $W_{r,a+x} \subset W_{r+1,a} \cap W_{g-1,b}$, and that $S \subset W_{r+1,a} \cap W_{g-1,b}$; the lemma is proved.

Proof of Torelli's theorem. Recall that we have identified $J(X)$ with $J(Y)$ ($= J$ say) and that V_r is the image of $S^r(Y)$ in J under A_Y.

Let r be the smallest integer ≥ 0 such that V_1 is contained in some translate of either W_{r+1} or W^*_{r+1}; since $V_1 \subset V_{g-1}$, and V_{g-1} is a translate of W_{g-1} by hypothesis, there is such an integer (e.g. $g-2$).

The theorem asserts that $r = 0$. Assume that $r \geq 1$, and that $V_1 \subset W_{r+1,a}$. Let $x \in W_1$, $y \in W_{g-1-r}$.

Remark that if x is fixed, there is an analytic set $Z(x) \subset W_{g-1-r}$, $Z(x) \neq W_{g-1-r}$, such that if $y \notin Z(x)$, then $V_1 \not\subset W_{g-1,b}$ where $b = a + x - y$.

In fact, suppose that $V_1 \subset W_{g-1,b}$ $\forall y$. Then $V_{1,-x-a} \subset \bigcap_{y \in W_{g-1-r}} W_{g-1,-y} = W_r$, contradicting the definition of r.

Now, if $V_1 \subset W_{g-1,a+x-y} = V_{g-1,\alpha-y}$, $\alpha = c_0 + x$, c_0 fixed, then $V_1 + y \subset V_{g-1,\alpha}$. By Lemma 1, this means that $y \in V_{g-2,\alpha}$. Hence, if $Z(x) = V_{g-2,\alpha} \cap W_{g-1-r} \neq W_{g-1-r}$, then, for $y \in W_{g-1-r} - Z(x)$, we have $V_1 \not\subset W_{g-1,b}$, $b = a + x - y$.

Consider now (with $b = a + x - y$, $y \notin Z(x)$)

$$V_1 \cap W_{g-1,b} = V_1 \cap (W_{g-1,b} \cap W_{r+1,a})\,;$$

since $V_1 \subset W_{r+1,a}$ and $V_1 \not\subset W_{g-1,b}$, we have $W_{r+1,a} \not\subset W_{g-1,b}$, so that by Lemma 3, we have

$$V_1 \cap W_{g-1,b} = (V_1 \cap W_{r,a+x}) \cup (V_1 \cap S)\,,$$

where $S = W_{r+1,a} \cap (W_{g-2,y-a})^*$. Since $V_1 \not\subset W_{g-1,b}$, (and W_{g-1} is a translate of V_{g-1}), there is a divisor $D(b)$ of degree g on Y such that (if $A_Y^{(g)}; S^g(Y) \to J$ is the natural map)

$$A_Y(D(b)) = V_1 \cdot W_{g-1,b}\,, \quad A_Y^{(g)}(D(b)) = b - c_1,\ c_1 \ \text{a constant}\,.$$

We write $D(b) = D_0(x) + D_1(x,y)$, where $D_0(x)$ consists of the part of $D(b)$ which maps into $V_1 \cap W_{r,a+x}$ under A_Y, and no point in $D_1(x,y)$ maps into $W_{r,a+x}$. We now claim that $D_0(x)$ has degree 1, i.e. that $D_0(x)$ consists of a single point which occurs with multiplicity one in $V_1 \cap W_{g-1,b}$.

First, suppose that $\deg D_0(x) \geq 2$. Then, if we fix x and let y run over $W_{g-1-r} - Z(x)$, the image of $D_1(x,y)$ in J would lie in a fixed translate $(V_{g-2})_{-A_Y(D_0(x))}$ of V_{g-2}. But the image of $D_1(x,y)$ is a fixed translate of $b - A_Y(D_0(x))$, hence is a fixed translate of $-y$ (depending on x). Since $Z(x) \neq W_{r-1-r}$, it follows that

$$W^*_{g-1-r} \subset V_{g-2,\beta}\ , \ \beta = \beta(x)\ .$$

Hence

$$\bigcap_{-v \in V_{g-2,\beta}} V_{g-1,v} \subset \bigcap_{-v \in W^*_{g-1-r}} W_{g-1,v+c} \quad (\text{if } V_{g-1} = W_{g-1} + c)\ .$$

By Lemma 1, the term on the left is a translate of V_1, while that on the right is a translate of W^*_r, contradicting the definition of r.

Thus $\deg D_0(x) \leq 1$.

Now, $\deg D_0(x) \geq 1$; if this were not the case, $D(b) = D_1(x,y)$ would have its support in a finite set depending only on y [viz the set in Y whose image under A_Y is $S \cap V_1$; $S \cap V_1$ is finite being contained in $V_1 \cap W_{g-1,b}$, $V_1 \not\subset W_{g-1,b}$]. But then $A_Y^{(g)}(D_1(x,y)) = a + x - y - c_1$ would be independent of x for x in some non-empty open set in W_1.

Thus, $\deg D_0(x) = 1$.

As remarked above, if $y \in W_{g-1-r} - Z(x)$, $D_1(x,y)$ has its support in a finite set depending only on y. Hence we can find infinitely many points $x_\nu \in W_1$ ($\nu \geq 1$) so that $D_1(x_\nu, y) = D_1(y)$ is independent of ν. Thus $A_Y(D_0(x_\nu)) = a + x_\nu - y - c_0 - A_Y(D_1(y))$, and

$$A_Y(D_0(x_\nu)) - A_Y(D_0(x_1)) = x_\nu - x_1\ ,\ \nu \geq 1\ .$$

Clearly, $A_Y(D_0(x_\nu)) - A_Y(D_0(x_1)) \in V_{1,t}$, $t = -A_Y(D_0(x_1))$, and $x_\nu - x_1 \in W_{1,-x_1}$. Thus, the curves $V_{1,t}$ and $W_{1,-x_1}$ intersect in infinitely many points, and so must be equal. This proves the theorem.

19. Riemann's Theorem on the Singularities of Θ

Riemann's singularity theorem expresses the order of vanishing of the ϑ-function at a point $\zeta \in \Theta$ in terms of $\dim |D|$, where $D \geq 0$ is a divisor of degree $g-1$ with $\zeta - \kappa = A(D)$. Riemann proves this by relating this order to the vanishing of ϑ on sets of the form $W_r - W_r - \zeta$ (*Über das Verschwinden der Theta-Functionen*).

The theorem has been formulated more geometrically (using the tangent cones to Θ) and generalised to W_k, $2 \leq k \leq g-1$, by G. Kempf: On the geometry of a theorem of Riemann, *Annals of Math.* 98 (1973), 178 – 185. For a discussion of this theorem and this whole circle of ideas, one cannot do better than consult the book of Arbarello–Cornalba–Griffiths-Harris [10].

We start with two lemmas which we have essentially proved before in connection with the Riemann factorisation theorem.

Lemma 1. *Given $P \in X$, there exists $\zeta \in \Theta$ such that the function* $x \mapsto \vartheta\big(A(x) - A(P) - \zeta\big)$ *is not* $\equiv 0$.

Proof. If $Y \subset S^g(X)$ is the set of critical points of the map $A : S^g(X) \to J(X)$, $Y = \{D \in S^g(X) \mid \text{rank of } dA \text{ at } D \text{ is } < g\}$, then $A|Y$ has no isolated points in any of its fibres (§15, Theorem 2) so that $Y' = A(Y)$ has dimension $\leq g-2$.

Now, if $x \mapsto \vartheta\big(A(x) - A(P) - \zeta\big) \equiv 0$, then $\zeta + A(P) = A(D) + \kappa$ where $D \geq 0$ has degree g and $\dim |D| > 0$, i.e. $D \in Y$ (§15, Theorem 2 again). Thus, we have only to choose $\zeta \in \Theta$, $\zeta \notin \kappa - A(P) + Y'$.

In what follows, we denote by $F_P (P \in X)$ the section $x \mapsto \vartheta\big(A(x) - A(P) - \zeta\big)$ (of a suitable line bundle on X).

Lemma 2. *Let $\zeta \in \Theta$; if P is such that $F_P(x) \not\equiv 0$, then* $\operatorname{div}(F_P) = P + D_0$, *where* $D_0 \geq 0$, $\deg D_0 = g-1$ *and D_0 is independent of P. (It can depend on ζ.)*

Proof. If $D = \operatorname{div}\big(\vartheta(A(x) - A(P) - \zeta)\big)$, then $D \geq 0$, $\deg D = g$ and $\dim |D| = 0$ (since $F_P \not\equiv 0$); moreover $F_P(P) = \vartheta(-\zeta) = \vartheta(\zeta) = 0$. Hence $D = P + D_0$, $D_0 \geq 0$, $\deg D_0 = g-1$ and $\dim |D_0| = 0$.

Let Q be such that $F_Q \not\equiv 0$; then

$$D' = \operatorname{div}(F_Q) = Q + D_1\,, \quad D_1 \geq 0\,, \deg D_1 = g-1\,.$$

We have $A(D) = A(P) + \zeta - \kappa$, so that $A(D_0) = \zeta - \kappa$; similarly, $A(D_1) = \zeta - \kappa$, and since D_0, D_1 have degree $g-1$, Abel's theorem implies that $D_0 \sim D_1$, so that, since $\dim |D_0| = 0$, we have $D_0 = D_1$.

Lemma 3. *Let $\zeta \in \Theta$, and suppose that there is $P \in X$ such that $F_P \not\equiv 0$. Then, there are at most g points $Q \in X$ with $F_Q \equiv 0$.*

Proof. Choose x_0 so that $\vartheta\big(A(x_0) - A(P) - \zeta\big) \neq 0$. The function $y \mapsto \vartheta\big(A(x_0) - A(y) - \zeta\big)$ is not $\equiv 0$, and its divisor is $\operatorname{div}\big(\vartheta(A(y) - \zeta')\big)$ (where $\zeta' = -\zeta + A(x_0)$), since the ϑ-function is even, so has degree g.

Theorem 1. *Let $\zeta \in \Theta$. Then $F_P \equiv 0\ \forall P \in X$ (i.e. $\vartheta\big(A(x) - A(P) - \zeta\big) = 0\ \forall x,\ \forall P$) if and only if*

$$\frac{\partial \vartheta}{\partial z_\nu}(\zeta) = 0 \quad \text{for} \quad \nu = 1, \ldots, g\,.$$

Proof. Suppose that $\vartheta\big(A(x) - A(P) - \zeta\big) = 0\ \forall x,\ \forall P$. If we differentiate with respect to x, we obtain, since $dA(x) = \big(\omega_1(x), \ldots, \omega_g(x)\big)$,

$$\sum_{\nu=1}^{g} \frac{\partial \vartheta}{\partial z_\nu}\big(A(x) - A(P) - \zeta\big)\omega_\nu(x) = 0\,;$$

setting $x = P$, we have

$$\sum_{\nu=1}^{g} \frac{\partial \vartheta}{\partial z_\nu}(-\zeta)\omega_\nu(P) = 0\ \forall P\,;$$

since $(\omega_1, \ldots, \omega_g)$ are linearly independent, and $\frac{\partial \vartheta}{\partial z_\nu}(z) = -\frac{\partial \vartheta}{\partial z_\nu}(-z)$, we have $\frac{\partial \vartheta}{\partial z_\nu}(\zeta) = 0$, $\nu = 1, \ldots, g$.

Suppose conversely that $x \mapsto \vartheta\big(A(x) - A(P) - \zeta\big)$ is $\not\equiv 0$; let $\operatorname{div}\big(\vartheta\big(A(x) - A(P) - \zeta\big)\big) = P + D_0(\zeta)$, $\deg D_0(\zeta) = g - 1$. By Lemma 3, there exists $Q \notin \operatorname{supp} D_0(\zeta)$ such that $\vartheta\big(A(x) - A(Q) - \zeta\big) \not\equiv 0$. By Lemma 2, $\operatorname{div}\big(\vartheta\big(A(x) - A(Q) - \zeta\big)\big) = Q + D_0(\zeta)$, so that Q is a *simple* zero of $\vartheta\big(A(x) - A(Q) - \zeta\big)$, so that

$$\sum_{\nu=1}^{g} \frac{\partial \vartheta}{\partial z_\nu}\big(A(x) - A(Q) - \zeta\big)\Big|_{x=Q} \omega_\nu(Q) \neq 0\,,$$

and $\frac{\partial \vartheta}{\partial z_\nu}(-\zeta) \neq 0$ for some ν.

This theorem can be formulated as follows. Recall that if E, E' are subsets of $J(X)$ and $\zeta \in J(X)$, then $E - E' - \zeta$ is the set $\{x - y - \zeta | x \in E, y \in E'\}$; we use similar notation $E + E'$ and so on.

Theorem 1'. *Let $\zeta \in \Theta$. Then ζ is a singular point of Θ if and only if $\vartheta(W_1 - W_1 - \zeta) \equiv 0$.*

Suppose now that the genus $g \geq 2$. Given $\zeta \in \Theta$, there is an integer $r < g$ such that $\vartheta(W_r - W_r - \zeta) \not\equiv 0$; in fact, since $W_{g-1} = A(K_X) - W_{g-1}$ (see proof of §17, Theorem 3), we see that $W_{g-1} - W_{g-1}$ is a translate of $W_{g-1} + W_{g-1} = J(X)$.

Given $\zeta \in \Theta$, let $r = r_\zeta$ be the largest integer such that $\vartheta(W_k - W_k - \zeta) \equiv 0$ for $k < r$; we have $r < g$.

Theorem 2. *Let s be an integer > 0, and $\zeta \in \Theta$. Then $r_\zeta \geq s$ if and only if $\zeta = \kappa + A(D)$, where $D \geq 0$, $\deg D = g - 1$ and $\dim |D| \geq s - 1$.*

Proof. Let $r = r_\zeta$. Then, there are effective divisors D_0, D_1 of degree r with $\vartheta\big(A(D_0) - A(D_1) - \zeta\big) \neq 0$. We may suppose that $\operatorname{supp}(D_0 + D_1)$ consists of $2r$ distinct points. Further, for fixed D_1, the set of $D_0 \in S^r(X)$ satisfying this condition is a non-empty open set in $S^r(X)$.

We write $D_0 = P + \Delta_0$ where $\deg \Delta_0 = r - 1$. The function F: $x \mapsto \vartheta\big(A(x) + A(\Delta_0) - A(D_1) - \zeta\big)$ is $\not\equiv 0$ (it is $\neq 0$ if $x = P$). If $x \in \operatorname{supp}(D_1)$, then $A(x) + A(\Delta_0) - A(D_1) = A(\Delta_0) - A(D_1 - x) \in W_{r-1} - W_{r-1}$. Hence, since $\vartheta(W_{r-1} - W_{r-1} - \zeta) = 0$, $F(x) = 0$ if $x \in \operatorname{supp}(D_1)$. Hence the divisor D of F, which is of degree g, can be written

$$D = D_1 + D_2 \,, \deg D_2 = g - r \,.$$

On the other hand,

$$A(D) = \zeta + A(D_1) - A(\Delta_0) - \kappa$$

and it follows that

$$\zeta - \kappa = A(D_2 + \Delta_0) \,, \deg(D_2 + \Delta_0) = g - r + r - 1 = g - 1 \,.$$

Further Δ_0 could be an arbitrary divisor in a non-empty open set in $S^{r-1}(X)$ [since $D_0 = P + \Delta_0$ could be anywhere in a non-empty open set in $S^r(X)$]. Hence $\dim |D_2 + \Delta_0| \geq r - 1$.

Conversely, suppose that $\zeta - \kappa = A(D)$, where $D \geq 0$, $\deg D = g-1$ and $\dim |D| \geq s-1$.

Given any effective divisor D_1 of degree $s - 1$, we may assume that $D \geq D_1$ (since $\dim |D| \geq s - 1$).

Let E_0, E_1 be effective divisors of degree $s - 1$. Choose $D \geq 0$, $\deg D = g - 1$, with $\zeta - \kappa = A(D)$ and $D \geq E_0$. We have

$$A(E_0) - A(E_1) - \zeta = A(E_0 - D) - A(E_1) - \kappa = -\big(\kappa + A\big(E_1 + (D - E_0)\big)\big) \,;$$

since $D - E_0 \geq 0$ and has degree $g - 1 - (s - 1)$, $E_1 + D - E_0 \geq 0$ and has degree $g - 1$; hence $\kappa + A(E_1 + D - E_0) \in W_{g-1} + \kappa = \Theta$, so that $A(E_0) - A(E_1) - \zeta \in -\Theta = \Theta$, and $\vartheta\big(A(E_0) - A(E_1) - \zeta\big) = 0$, i.e. $\vartheta(W_{s-1} - W_{s-1} - \zeta) \equiv 0$.

Theorem 3. (Riemann's singularity theorem). *Let $\zeta \in \Theta$ and let m be the order of vanishing of ϑ at ζ, i.e. if $\alpha = (\alpha_1, \ldots, \alpha_g)$ and $|\alpha| = \alpha_1 + \ldots + \alpha_g < m$, then*

$$\frac{\partial^\alpha \vartheta}{\partial z^\alpha}(\zeta) = \frac{\partial^{|\alpha|} \vartheta}{\partial z_1^{\alpha_1} \ldots \partial z_g^{\alpha_g}}(\zeta) = 0 \,,$$

while there is β with $|\beta| = m$ so that $\frac{\partial^\beta \vartheta}{\partial z^\beta}(\zeta) \neq 0$.

Equivalently, the expansion of ϑ in a series of homogeneous polynomials in $z - \zeta$ starts with a homogeneous polynomial of degree m.

Then $m = r_\zeta$, so that $m = 1 + \dim|D|$, where $D \geq 0$ is a divisor of degree $g - 1$ such that $\zeta - \kappa = A(D)$.

Proof. Suppose that $k \geq 1$ is an integer such that $\vartheta(W_k - W_k - \zeta) \equiv 0$. We claim that if $|\alpha| \leq k$, we have $\frac{\partial^\alpha \vartheta}{\partial z^\alpha} \mid W_{k-|\alpha|} - \zeta \equiv 0$. We prove this by induction on $|\alpha| = n$. If $n = 0$, this is just our hypothesis. Assume that the assertion is proved for $|\alpha| = n-1 < k$. Then, if $u, v \in W_{k-n}$ and $x, y \in X$, we have $A(x)+u-\big(A(y)+v\big)-\zeta \in W_{k-(n-1)}-W_{k-(n-1)}-\zeta$, so that $\frac{\partial^\alpha \vartheta}{\partial z^\alpha}\big(A(x) + u - A(y) - u - \zeta\big) = 0$. Differentiating this with respect to x and setting $x = y$, we obtain

$$\sum_{\nu=1}^{g} \frac{\partial}{\partial z_\nu} \frac{\partial^\alpha \vartheta}{\partial z^\alpha}(u - v - \zeta)\omega_\nu(x) = 0 \quad \forall x \in X ,$$

so that $\frac{\partial^{\alpha'} \vartheta}{\partial z^{\alpha'}}(u - v - \zeta) = 0$ if $|\alpha'| = |\alpha| + 1 = n$ and all $u, v \in W_{k-n}$.

This implies that $\frac{\partial^\alpha \vartheta}{\partial z^\alpha}(-\zeta) = 0$ for $|\alpha| \leq k$, so that we have $m \geq r_\zeta$.

The inequality $m \leq r_\zeta$ is harder to prove. Let $r = r_\zeta$, so that $\vartheta(W_{r-1} - W_{r-1} - \zeta) \equiv 0$, and $\vartheta(W_r - W_r - \zeta) \not\equiv 0$. We can find three effective divisors D, D_0, D_1 of degree r such that $\operatorname{supp}(D + D_0 + D_1)$ consists of $3r$ distinct points and such that

$$\vartheta\big(A(D_0) - A(D_1) - \zeta\big) \neq 0 , \quad \vartheta\big(A(D_0) - A(D) - \zeta\big) \neq 0 .$$

Let $D = \sum_{\nu=1}^r P_\nu$, $D_1 = \sum_{\nu=1}^r Q_\nu$ and let $\omega_{P_\nu Q_\nu}$ be the normalized abelian differential of the third kind (residue $+1$ at P_ν, -1 at Q_ν, holomorphic on $X - \{P_\nu, Q_\nu\}$ and having a-periods 0). Let $\varphi = \sum_{\nu=1}^r \omega_{P_\nu Q_\nu}$.

For $x_1, \ldots, x_r \in X$, set

$$F(x_1, \ldots, x_r) = \exp\Big(\sum_{\nu=1}^{r} \int_{P_0}^{x_\nu} \varphi\Big) \frac{\vartheta\big(A(x_1) + \cdots + A(x_r) - A(D_1) - \zeta\big)}{\vartheta\big(A(x_1) + \cdots + A(x_r) - A(D) - \zeta\big)} .$$

(F is holomorphic and $\neq 0$ at $(p_1, \ldots, p_r)$ if $\Sigma p_\nu = D_0$.)

We claim that F defines a meromorphic function on $X \times \cdots \times X$ (r times). To see this, we fix $x_2, \ldots, x_r$ and consider F as a function $F(x_1)$ defined on the simply connected polygon Δ obtained by slitting X along the homology basis a_i, b_j.

If $x_1 \in b_j$ and x_1' is the corresponding point of b_j', we have $A(x_1) = A(x_1') + e_j$ and $\int_{x_1'}^{x_1} \varphi = 0$ since the a-periods of φ are 0. Hence $F(x_1) = F(x_1')$. If $x_1 \in a_j$ and x_1' is the corresponding point in a_j', we have $A(x_1) - A(x_1') = -B_j$ and, by the reciprocity theorem in §14,

$$\int_{x_1'}^{x_1} \varphi = -\sum_{\nu=1}^{r} \int_{b_j} \omega_{P_\nu Q_\nu} = 2\pi i \sum_{\nu=1}^{r} \int_{P_\nu}^{Q_\nu} \omega_j = 2\pi i \sum_{\nu=1}^{r} \big(A_j(Q_\nu) - A_j(P_\nu)\big)$$

so that

$$\frac{F(x_1)}{F(x_1')} = e^{2\pi i\left(A_j(D_1)-A_j(D)\right)} \frac{\exp\big(2\pi i(\sum_\nu A_j(x_\nu) - A_j(D_1) - \zeta_j) + \pi i\, B_{jj}\big)}{\exp\big(2\pi i(\sum_\nu A_j(x_\nu) - A_j(D) - \zeta_j) + \pi i\, B_{jj}\big)} = 1\,.$$

Thus $F(x_1)$ is single-valued on X.

We now claim that F is constant ($\neq 0$ since $F(p_1, \dots, p_n) \neq 0$). Again, fix $x_2, \dots, x_n$ generically in X and consider the divisors

$$\begin{aligned} E_1 &= \operatorname{div}\big(x_1 \mapsto \vartheta(A(x_1) + \dots + A(x_r) - A(D_1) - \zeta)\big)\,, \\ E &= \operatorname{div}\big(x_1 \mapsto \vartheta(A(x_1) + \dots + A(x_r) - A(D) - \zeta)\big)\,. \end{aligned}$$

If $x_1 \in \operatorname{supp}(D_1)$, we have $\sum_1^r A(x_\nu) - A(D_1) - \zeta = \sum_2^r A(x_\nu) - A(D_1 - x_1) - \zeta \in W_{r-1} - W_{r-1} - \zeta$, so that ϑ vanishes at this point. Hence $E_1 = D_1 + D_1'$, where $D_1' \geq 0$ and has degree $g-r$. Similarly, $E = D + D'$, where $D' \geq 0$ and $\deg D' = \deg D_1' = g-r$.

We now prove (using the same reasoning as in Lemma 2) that $D_1' = D'$. In fact, we have $\dim|E| = 0$ (since $\vartheta(A(x_1) + \dots + A(x_r) - A(D) - \zeta) \not\equiv 0$ in x_1) so that $\dim|D'| = 0$. Moreover, $A(D') = A(E) - A(D) = \big(\zeta + A(D) - \sum_2^r A(x_\nu) - \kappa\big) - A(D) = \zeta - \kappa - \sum_2^r A(x_\nu)$; in the same way, $A(D_1') = \zeta - \kappa - \sum_2^r A(x_\nu)$, so that, since $\deg D' = \deg D_1'$, Abel's theorem implies that $D_1' \sim D'$. Since $\dim|D'| = 0$, we have $D' = D_1'$.

Consider now the function $\exp\big(\sum_{k=1}^r \int_{P_0}^{x_k} \varphi\big)$ with $x_2, \dots, x_r$ fixed. If z_ν, w_ν are local coordinates at P_ν, Q_ν respectively (with $z_\nu(P_\nu) = 0, w_\nu(Q_\nu) = 0$), then if x_1 is near P_ν, we have $\int_{P_0}^{x_1} \varphi = \sum_{k=1}^r \int_{P_0}^{x_1} \omega_{P_k Q_k} = \int_{P_0}^{x_1} \frac{dz_\nu}{z_\nu} + h$, h holomorphic at P_ν while, if x_1 is near Q_ν, we have $\int_{P_0}^{x_1} \varphi = -\int_{P_0}^{x_1} \frac{dw_\nu}{w_\nu} + h'$, h' holomorphic at Q_ν. Hence near P_ν,

$$\exp\Big(\sum_{k=1}^{r} \int_{P_0}^{x_k} \varphi\Big) = c(x_2, \dots, x_r) z_\nu e^h \quad (c(x_2, \dots, x_r) \in \mathbb{C} - \{0\})$$

and, near Q_ν,

$$\exp\Big(\sum_{k=1}^{r} \int_{P_0}^{x_k} \varphi\Big) = c'(x_2, \dots, x_r) w_\nu^{-1} e^{h'} \quad (c'(x_2, \dots, x_r) \in \mathbb{C} - \{0\})\,.$$

Consequently, the divisor of $\exp\big(\sum_{k=1}^r \int_{P_0}^{x_k} \varphi\big)$, as a function of x_1, is $\Sigma P_\nu - \Sigma Q_\nu = D - D_1$. Hence, the divisor of $x_1 \mapsto F(x_1, x_2, \dots, x_r)$ is

$$D - D_1 + E_1 - E = D_1' - D' = 0\,.$$

Thus, this function is holomorphic and non-zero on X, and so is a constant $\neq 0$ (for fixed generic $x_2, \ldots, x_r$). From the symmetry of the function in $x_1, \ldots, x_r$, it follows that $F \equiv c_0 \in \mathbb{C} - \{0\}$.

We have proved: There is a constant $c_0 \neq 0$ so that

$$\exp\Big(\sum_{\nu=1}^{r} \int_{P_0}^{x_\nu} \varphi\Big) \vartheta\big(A(x_1)+\cdots+A(x_r)-A(D_1)-\zeta\big) = c_0 \vartheta\big(A(x_1)+\cdots+A(x_r)-A(D)-\zeta\big) .$$

We differentiate with respect to x_1 and set $x_1 = P_1$. This gives, since $\exp\big(\int_{P_0}^{x_1} \varphi\big) = 0$ when $x_1 = P_1$ (see above)

$$\begin{aligned} &c_1 \exp\Big(\sum_{\nu=2}^{r} \int_{P_0}^{x_\nu} \varphi\Big) \vartheta\big(A(P_1) + A(x_2) + \ldots + A(x_r) - A(D_1) - \zeta\big) dz_1(P_1) \\ &= c_0 \sum_{k=1}^{g} \frac{\partial \vartheta}{\partial z_k} \big(A(P_1) + A(x_2) + \ldots + A(x_r) - A(D) - \zeta\big) \omega_k(P_1) ; \end{aligned}$$

here $z_1(z_\nu)$ is a local coordinate at $P_1(P_\nu)$ and $c_1 \neq 0$; in fact, we saw above that $\exp\big(\int_{P_0}^{x_1} \varphi\big) = z_\nu e^h$, h holomorphic at P_1, in a neighbourhood of P_1. Iterating this argument, we find that

$$\begin{aligned} &c_1 \ldots c_r \, \vartheta\big(A(P_1) + \ldots + A(P_r) - A(D_1) - \zeta\big) \, dz_1(P_1) \ldots dz_r(P_r) \\ &= c_0 \sum_{1 \leq k_1, \ldots, k_r \leq g} \frac{\partial^r \vartheta}{\partial z_{k_1} \ldots \partial z_{k_r}} \big(A(P_1) + \ldots + A(P_r) - A(D) - \zeta\big) \omega_{k_1}(P_1) \ldots \omega_{k_r}(P_r) . \end{aligned}$$

Since the term on the left is $\neq 0$ and $A(P_1) + \ldots + A(P_r) = D$, it follows that there exist $k_1, \ldots, k_r$ (between 1 and g) so that

$$\frac{\partial^r \vartheta}{\partial z_{k_1} \ldots \partial z_{k_r}} (-\zeta) \neq 0 .$$

This proves that $m \leq r = r_\zeta$, and with it, Riemann's theorem.

In presenting essentially this argument in [3], Riemann does not use the normalized abelian differentials $\omega_{P_\nu Q_\nu}$ explicitly, but rather, their expression in terms of the ϑ-function given by §17, Theorem 6. Thus, we could have worked directly with the function

$$F(x_1, \ldots, x_r) = \frac{\prod_{k,\ell=1}^{r} \vartheta\big(A(x_k) - A(P_\ell) - \zeta_0\big)}{\prod_{k,\ell=1}^{r} \vartheta\big(A(x_k) - A(Q_\ell) - \zeta_0\big)} \, \frac{\vartheta\big(\sum_{\nu=1}^{r} A(x_\nu) - A(D_1) - \zeta\big)}{\vartheta\big(\sum_{\nu=1}^{r} A(x_\nu) - A(D) - \zeta\big)} ,$$

where ζ_0 is a general point of Θ. The argument, using this function, remains essentially the same.

One of the applications of the singularity theorem is to the proof that the set Θ_{sing} of singular points of Θ has pure dimension $g-4$ if X is not hyperelliptic (and dimension $g-3$ if X is hyperelliptic). This is an important property, and shows that Jacobians are special among the so called principally polarised abelian varieties (complex tori defined by a lattice having a basis (I, B) where I is the identity, and B a complex symmetric matrix with positive definite imaginary part). The theta divisor in the generic case (defined as on the Jacobian) is smooth. We shall not prove this theorem, but shall say a few words about why $\Theta_{\text{sing}} \neq \emptyset$ for $g \geq 4$. The theorem is discussed in [8], [10], as well as a partial converse studied by A. Andreotti and A. Mayer [On period relations for abelian integrals on algebraic curves, Annali Sc. Norm. Pisa 21 (1967), 189 – 238.].

Because of §19, Theorem 1', Θ_{sing} is a translate of the set $W^1_{g-1} = \{$ Divisors $D \geq 0$ of degree $g-1$ with $\dim |D| > 0\}$. Thus, the statement $\Theta_{\text{sing}} \neq \emptyset$ is equivalent to the statement that there is a non-constant-meromorphic function on X whose divisor of poles has degree $\leq g-1$. If X is hyperelliptic, this is obvious when $g \geq 3$. One way to prove this when X is not hyperelliptic (and $g \geq 4$) is to prove the following:

Let $X \subset \mathbb{P}^{g-1}$ be the canonical imbedding. Then, there is a quadric $Q = \sum_{1 \leq i,j \leq g} a_{ij} z_i z_j$ (with (a_{ij}) symmetric) such that $Q|X \equiv 0$ and $0 < \operatorname{rank}(a_{ij}) \leq 4$.

[This can actually be done by an easy dimension count, using Noether's theorem given in §13. Of course, the result is trivial if $g = 4$.]

Given this theorem, one can proceed as follows. Since a symmetric matrix can be diagonalised, we can make a linear transformation of $(z_1, \ldots, z_g)$ to put (a_{ij}) in the form

$$\begin{pmatrix} 1 & & & & & 0 \\ & \ddots & & & & \\ & & 1 & & & \\ & & & 0 & & \\ & & & & \ddots & \\ 0 & & & & & 0 \end{pmatrix}$$

and the rank condition means that there are at most four 1's on the diagonal, so that $Q = \sum_1^r z_\nu^2$, $r \leq 4$. If $r \leq 2$, Q is a product of linear forms, so cannot vanish on X since X is non-degenerate (not contained in any hyperplane). Thus $Q = z_1^2 + z_2^2 + z_3^2$ or $\sum_1^4 z_\nu^2$. Since $a^2 + b^2 = (a+ib)(a-ib)$, we can, by another linear transformation, assume that $Q = z_3^2 + z_1 z_2$ or $Q = z_1 z_2 + z_3 z_4$. If $(\omega_1, \ldots, \omega_g)$ is the basis of $H^0(X, \Omega)$ corresponding to this form of Q, the relation becomes

$$\omega_3^2 + \omega_1\omega_2 = 0 \quad \text{or} \quad \omega_1\omega_2 + \omega_3\omega_4 = 0$$

(as section of $K_X^{\otimes 2}$). In either case, if $\operatorname{div}(\omega_3) = \sum_{\nu=1}^{2g-2} P_\nu$, ω_1 or ω_2 (say ω_1) must vanish on at least $g-1$ of the points P_ν, and ω_1/ω_3 then has a polar divisor of degree $\leq g-1$.

The study of quadrics containing the canonical curve is a very rich and beautiful one. See, in particular

B. Saint–Donat. On Petri's analysis of quadrics through a canonical curve, *Math. Annalen* 206 (1973), 157 - 175.

M. Green. Quadrics of rank four in the ideal of the canonical curve. *Inv. Math.* 75 (1984), 85 - 104.

References

The literature on Riemann surfaces and the topics dealt with in these notes is vast, and we content ourselves with just a few references.

The classic book on Riemann surfaces is:

[1] H. Weyl. *Die Idee der Riemannschen Fläche*, Teubner 1913.

Riemann's two papers, which form the basis of much of the material in these notes are:

[2] B. Riemann. Theorie der Abel'schen Functionen. *J. für die reine und angew. Math.* 54 (1857). Collected Works: pp. 88 – 144.

[3] B. Riemann. Über das Verschwinden der Theta-Functionen. *J. für die reine und angew. Math.* 65 (1865). Collected Works: 212 – 224.

Two texts on Riemann surfaces which are accessible and have much in common with the first part of these notes are:

[4] O. Forster. *Riemannsche Flächen*, Springer 1977. There is also an english translation available.

[5] R.C. Gunning. *Lectures on Riemann surfaces.* Princeton Mathematical Notes, 1966.

For the topology of orientable surfaces, in particular the classification theorem, one can consult:

[6] W.S. Massey. *Algebraic Topology: An Introduction.* Harcourt Brace, New York, 1967.

For the standard material from complex analysis (in particular properties of $\bar{\partial}$) used in the first part of these notes, as well as a different arrangement of the proof of the finiteness theorem, see:

[7] R. Narasimhan. *Complex Analysis in one Variable*, Birkhäuser, 1985.

As a quick introduction to the many aspects of the geometry of curves and Jacobians, one cannot recommend too strongly the following beautiful book:

[8] D. Mumford. *Curves and their Jacobians.* University of Michigan Press, 1975.

Two other indispensable books dealing not only with the material of these notes, but with very much more are:

[9] P.A. Griffiths and J. Harris. *Principles of Algebraic Geometry.* Wiley, New York, 1978.

[10] E. Arbarello, M. Cornalba, P.A. Griffiths and J. Harris. *Geometry of Algebraic Curves*, Vol. I, Springer, 1985.

Serre's paper on the duality theorem is:

[11] J.-P. Serre. Un théorème de dualité. *Comm. Math. Helv.* 29 (1955), 2 – 26.

Martens' proof of Torelli's theorem is in:

[12] H. Martens. A new proof of Torelli's theorem. *Annals of Math.* 78 (1963), 107 – 111.

LM – Lectures in Mathematics, ETH Zürich

Department of Mathematics
Research Institute of Mathematics

Each year the Eidgenössische Technische Hochschule (ETH) at Zürich invites a selected group of mathematicians to give postgraduate seminars in various areas of pure and applied mathematics. These seminars are directed to an audience of many levels and backgrounds. Now some of the most successful lectures are being published for a wider audience through the ***Lectures in Mathematics, ETH Zürich*** *series. Lively and informal in style, moderate in size and price, these books will appeal to professionals and students alike, bringing a quick understanding of some important areas of current research.*

C. de Boor
Splinefunktionen
1990. ISBN 3-7643-2514-3

J.D. Monk
Cardinal Functions on Boolean Algebras
1990. ISBN 3-7643-2495-3

D. Bättig/H. Knörrer
Singularitäten
1991. ISBN 3-7643-2616-6

R.J. LeVeque
Numerical Methods for Conservation Laws
2nd Edition, 3rd Printing 1994.
1992. ISBN 3-7643-2723-5

R. Narasimhan
Compact Riemann Surfaces
1992. ISBN 3-7643-2742-1

A.J. Tromba
Teichmüller Theory in Riemannian Geometry
1992. ISBN 3-7643-2735-9

M. Yor
Some Aspects of Brownian Motion
1992. ISBN 3-7643-2807-X

G. Baumslag
Topics in Combinatorial Group Theory
1993. ISBN 3-7643-2921-1

M. Giaquinta
Introduction to Regularity Theory for Nonlinear Elliptic Systems
1993. ISBN 3-7643-2879-7

O. Nevanlinna
Convergence of Iterations for Linear Equations
1993. ISBN 3-7643-2865-7

R.-P. Holzapfel
The Ball and Some Hilbert Problems
1995. ISBN 3-7643-2835-5

J.F. Carlson
Modules and Group Algebras
Notes by Ruedi Suter
1996. ISBN 3-7643-5389-9

DMV Seminar

Workshops, edited by the German Mathematics Society

The workshops organized by the Gesellschaft für mathematische Forschung in cooperation with the Deutsche Mathematiker–Vereinigung (German Mathematics Society) are primarily intended to introduce students and young mathematicians to current fields of research. By means of these well-organized seminars, scientists from other fields will also be introduced to new mathematical ideas. The publication of these workshops proceedings in the **DMV-Seminar** series will make the material available to an ever larger audience.

DMV 1 *M. Knebusch / W. Scharlau:* Algebraic Theory of Quadratic Forms (ISBN 3-7643-1206-8)

DMV 2 *K. Diederich / I. Lieb:* Konvexität in der komplexen Analysis (ISBN 3-7643-1207-6)

DMV 3 *S. Kobayashi / H.-H. Wu:* Complex Differential Geometry (ISBN 3-7643-1494-X)

DMV 4 *R. Lazarsfeld / A. van de Ven:* Topics in the Geometry of Projective Space (ISBN 3-7643-1660-8)

DMV 5 *W.M. Schmidt:* Analytische Methoden für diophantische Gleichungen (ISBN 3-7643-1661-6)

DMV 6 *A. Delgado / D. Goldschmidt / B. Stellmacher:* Groups and Graphs: New Results and Methods (ISBN 3-7643-1736-1)

DMV 7 *R. Hardt / L. Simon:* Seminar on Geometric Measure Theory (ISBN 3-7643-1815-5)

DMV 8 *Y.-T. Siu:* Lectures on Hermitian-Einstein Metric for Stable Bundles and Kähler-Einstein Metrics (ISBN 3-7643-1931-3)

DMV 9 *P. Gaenssler / W. Stute:* Seminar on Empirical Processes (ISBN 3-7643-1921-6)

DMV 10 *J. Jost:* Nonlinear Methods in Riemannian and Kählerian Geometry Delivered at the German Mathematical Society Seminar in Düsseldorf in June, 1986 (ISBN 3-7643-2685-9)

DMV 11 *T. tom Dieck / I. Hambleton:* Surgery Theory and Geometry of Representations (ISBN 3-7643-2204-7)

DMV 12 *J.H. van Lint / G. van der Geer:* Introduction to Coding Theory and Algebraic Geometry (ISBN 3-7643-2230-6)

DMV 13 *H. Kraft / P. Slodowy / T.A. Springer* (Eds): Algebraische Transformationsgruppen und Invariantentheorie / Algebraic Transformation Groups and Invariant Theory (ISBN 3-7643-2284-5)

DMV 14 *R. Bhattacharya / M. Denker:* Asymptotic Statistics (ISBN 3-7643-2282-9)

DMV 15 *A.H. Schatz / V. Thomée / W.L. Wendland:* Mathematical Theory of Finite and Boundary Element Methods (ISBN 3-7643-2211-X)

DMV 16 *J. Mecke / R.G. Schneider / D. Stoyan / W.R.R. Weil:* Stochastische Geometrie (ISBN 3-7643-2543-7)

DMV 17 *L. Ljung / G. Pflug / H. Walk:* Stochastic Approximation and Optimization of Random Systems (ISBN 3-7643-2733-2)

DMV 18 *K.W. Roggenkamp / M.J. Taylor:* Group Rings and Class Groups (ISBN 3-7643-2734-0)

DMV 19 *P. Groeneboom / J.A. Wellner:* Information Bounds and Nonparametric Maximum Likelihood Estimation (ISBN 3-7643-2794-4)

DMV 20 *H. Esnault / E. Viehweg:* Lectures on Vanishing Theorems (ISBN 3-7643-2822-3)

DMV 21 *M.E. Pohst:* Computational Algebraic Number Theory (ISBN 3-7643-2913-0)

DMV 22 *H.W. Knobloch / A. Isidori / D. Flockerzi:* Topics in Control Theory (ISBN 3-7643-2953-X)

DMV 23 *M. Falk / J. Hüsler / R.-D. Reiss:* Laws of Small Numbers: Extremes and Rare Events 3 1/2" Diskette included (ISBN 3-7643-5071-7)

DMV 24 *M. Aubry:* Homotopy Theory and Models. Based on lectures held at a DMV Seminar in Blaubeuren by H.J. Baues, S. Halperin and J.-M. Lemaire (ISBN 7643-5185-3)

DMV 25 *W. Ballmann:* Lectures on Spaces of Nonpositive Curvature (ISBN 3-7643-5242-6)